The Nature of War

ALSO BY JIM STEMPEL

*The Battle of Glendale: The Day the South
Nearly Won the Civil War* (McFarland, 2011)

The CSS Albemarle *and William Cushing:
The Remarkable Confederate Ironclad
and the Union Officer Who Sank It* (McFarland, 2011)

The Nature of War
Origins and Evolution of Violent Conflict

JIM STEMPEL

McFarland & Company, Inc., Publishers
Jefferson, North Carolina, and London

LIBRARY OF CONGRESS CATALOGUING-IN-PUBLICATION DATA

Stempel, Jim, 1948–
  The nature of war : origins and evolution of violent conflict / Jim Stempel.
      p.   cm.
  Includes bibliographical references and index.

  ISBN 978-0-7864-6837-9
  softcover : acid free paper ∞

  1. War. 2. War — Causes. 3. War — Psychology.   I. Title.
  U21.2.S698 2012
  355.02 — dc23   2012031702

BRITISH LIBRARY CATALOGUING DATA ARE AVAILABLE

© 2012 Jim Stempel. All rights reserved

No part of this book may be reproduced or transmitted in any form or by any means, electronic or mechanical, including photocopying or recording, or by any information storage and retrieval system, without permission in writing from the publisher.

Front cover images © 2012 Shutterstock

Manufactured in the United States of America

McFarland & Company, Inc., Publishers
  Box 611, Jefferson, North Carolina 28640
    www.mcfarlandpub.com

Table of Contents

Preface 1
Prologue 3

One. Some Proposals on the Origin of War 11
Two. The Developmental Path 23
Three. Civilization's Dawn 35
Four. Civilization, Eden and the Romantic Blunder 43
Five. The Psychology of War 50
Six. The Archetype, the Ego and the Genesis of War 61
Seven. Alexander and the Warrior Archetype 72
Eight. A New Way Emerges 83
Nine. Revolution 92
Ten. A Violent Clash of Levels 112
Eleven. Documents of Change 125
Twelve. A War to End All Wars 137
Thirteen. The Great War, Act II 151
Fourteen. Cold War Conflict 162
Fifteen. The Evolution of War 170
Sixteen. The Future's Promise 177
Seventeen. Understanding Weapons 185
Eighteen. Peace 191

Chapter Notes 199
Bibliography 207
Index 211

War is
A grave affair of state;
It is a place
Of life and death,
A road
To survival and extinction,
A matter
To be pondered carefully.
　　　　— Sun–tzu (*The Art of War*)

Preface

From where has war come, and what drives its persistence? That is the subject of this book. The question is a simple one, but the answer has proven elusive. Everyone living today is of the subspecies *Homo sapiens sapiens*—the wisest of the wise the title means, roughly speaking—yet for almost 12,000 years we have slaughtered one another with such persistence, brutality, and enterprise that the designation seems almost incomprehensible. Is it satire? In the twentieth century alone our wars claimed over ninety million lives. Does that have the ring of wisdom about it?

If, in fact, humankind were as much a product of mind as most of us would like to believe, it would seem that the great brain which we are told separates humans from the less cerebral but nonwarring primates should long ago have disabused our ancestors of the urge to war. Cooperation and compromise, after all, is clearly the most sensible, rational path to longevity, prosperity, and human progress; but that nonviolent path, by and large, has not been the path our subspecies has chosen. Since roughly the time of recorded history war has been our way, our path, the organizational principal upon which much human endeavor has been founded. Why?

In *A History of Warfare* historian John Keegan observes: "warfare is almost as old as man himself, and reaches into the most secret places of the human heart, places where self dissolves rational purpose, where pride reigns, where emotion is paramount, where instinct is king."[1] If war is so base, after all—so instinctive, such a primitive aspect of the human chronicle as Keegan suggests—then why, it seems fair to ask, is it not also more fundamentally comprehensible? Why is warfare "almost" as old as man himself, and why does war touch upon such dark, secret places in the human heart? How is it that a creature capable of producing great art, architecture, literature, medicine, and wondrous acts of compassion is simultaneously capable of such cruel and wanton slaughter? The answers to these questions are not easy, and they will require a journey across time and disciplines to solve, from C.G. Jung to Adolph Hitler, from Sumer to Pearl Harbor, from military science to psychology — and these are just a few of the names, places, and disciplines this inquiry will touch upon. Yet in the end the journey will prove well worth the effort: war understood, the human journey placed into a more meaningful context, peace made a word worth remembering.

My previous research and writings have been focused upon human development, history, and warfare. This at first might seem an odd, disjointed combination, but it has always been a fundamental belief of mine that the three are inexorably linked, that humanity cannot be properly understood without somehow making sense of the extremes. Art, architecture, and Austerlitz: can all three somehow be made sense of? The answer is yes.

For some 12,000 years human beings have warred. From Jericho to Hiroshima, our trail has been one of slaughter and annihilation; no culture has been left untouched, no stone left unturned. From spear point to thermonuclear ordnance, our ingenuity has also proven our downfall. Why is this so? To say that human emergence has been awkward, violent, and destructive is to say the least, yet somehow *Homo sapiens sapiens* has managed to survive, emerging from twelve millennia of self-inflicted blood and ruin into a twenty-first century that promises to be our best, our worst, or quite possibly our last.

The wisest of the wise: thus have we named ourselves. Only time will tell if this be fact or satire. We will either come to grips with ourselves and the cause of our wars — our endless, irrational hatreds — or we will perish in a steadily burning fireball of our worst fears. I firmly believe that with understanding and modest development humanity will rise to the occasion, that we are capable of understanding war and therefore of leaving it behind just as maturity discards its youth. The converging nature of our world and the simultaneous proliferation of our most lethal ordnance make this leap imperative. Indeed, the choice now before humankind appears to be one of wisdom or of potential extinction. We have not always opted for wisdom. Perhaps its time has come.

Prologue

The story of war begins on a long-ago landscape — distant, rolling, beautiful. Imagine that the sky above this landscape is crystal blue and that vast fields of wild grains stretch before your eyes for mile upon mile, rippling in a soft breeze like waves across an immense ocean. The color of these grains is amber, the hills brown, the bordering woods green, the tumbling creeks that snake across this fresh landscape a sparkling blue. Sketch in a few grazing herds of wild sheep, gazelles, cattle, and, finally, a warm Mediterranean breeze that gently tosses the treetops and the scene is complete.

This is the landscape of the Fertile Crescent — "the horseshoe-shaped sweep of land stretching around the Arabian Desert, from the Zagros Mountains of Iran through Northern Iraq, Syria, and southern Turkey to Lebanon and Israel"[1] — as it might well have looked some twelve thousand years ago. It is open and pleasant. Humans are few. Just a few thousand years before, the great ice caps that formed the ice age glaciers and peaked around 16,000 B.C. had begun to recede, allowing vegetation to again spread across the globe and the human population to expand rapidly. By 10,000 B.C. that population had grown to around 8 million worldwide,[2] and the Fertile Crescent began to teem with small, roving bands of about thirty to forty people. These were bands or tribes of hunter-gatherers, all consisting of the subspecies of human known as *Homo sapiens sapiens*,[3] from which all modern humans have since emerged. In appearance they looked distinctly contemporary and today, if properly attired, might pass unnoticed on the streets of any modern city.

According to the most current anthropological studies, *Homo Sapiens sapiens* evolved in Africa sometime around 130,000 B.C., then spread across the globe, rapidly supplanting earlier, less robust hominid strains such as Neanderthal (*Homo sapiens neanderthalensis*).[4] *Homo sapiens sapiens* inherited the stone weapons of previous protohuman generations but invented the bow and arrow,[5] learned to hunt in teams, built reasonably elaborate shelters out of wood, dirt, and animal skins, and devised the first sophisticated hunting strategies. Despite these advances it was a dangerous, difficult life.

John Keegan explains: Those were violent times, as were the hundreds of thousands of years in which man pitted himself against large animals. At Arene Candide, in Italy, has been found the skeleton of a young man who died at the end of the Old Stone Age, at least 10,000 years ago. Part of his lower jaw, his collar-bone and the shoulder-blade, together with the top of the thigh bone, had been carried away by the bites of a large, savage animal, perhaps a bear which had been cornered in a pit or cave that the hunters dug or adapted as a trap.[6]

Despite the harsh world and brutal hardships associated with Old Stone Age life, *Homo sapiens sapiens* demonstrated a remarkable capacity for learning, cooperation, and invention. While human and protohuman beings (going back to *Homo erectus*) had been in existence for some two million years, the earth had never seen anything like *Homo sapiens sapiens* before. Yet, as prehistorians note, while these hunters may have appeared modern, they were not far removed either intellectually or psychologically from the animals they hunted; there was no gulf, no

> great abyss separating [him] from the animal. The bonds between them were not yet broken, and man still felt near the beasts that lived around him, that killed and fed like him.... From them he still retained all the faculties that civilization has blunted — rapid action and highly trained senses of sight, hearing and smell, physical toughness in an extreme degree, a detailed, precise knowledge of the qualities and habits of game, and great skill in using with the greatest effect the rudimentary weapons available.[7]

While *Homo sapiens sapiens* was of necessity an aggressive mammal, and while occasional fighting between roving bands of hunters unquestionably took place, all archeological evidence — excavations, cave paintings, artifacts, and the like — indicate that war as we now define that term did not yet exist. Thomas Hobbes' famous quote about the life of primitive humankind as being "solitary, poor, nasty, brutish, and short"[8] was undoubtedly correct in many regards, but that life appears to have been none of those things due to the vagaries of war. Robert O'Connell, for instance, in his examination of weapons points out that the notion of men engaging in war during this period is unsupportable. "If his life was dangerous," says O'Connell, "it was probably more due to hunting than fighting."[9] Or as Stanton Coblentz in his study of war puts it, "Everything we have learned of these 'children of nature' makes it reasonable to assume that the Old Stone Age was actually a time of peace."[10]

John Keegan in his *A History of Warfare* states, "If we are to think of the bowmen of the New Stone Age as prototypical of the hunters who still survive in the modern world, it is certainly not safe to invest them with strong warrior qualities; it is equally unsafe to argue that they were peaceful people."[11] It seems probable, therefore, that hunters of the Old Stone Age were aggressive, and at times clashed with other hunting parties — or perhaps even with one another in their own party from time-to-time — but they were not yet predisposed to war. Robert O'Connell seems to share that same view, noting the aggressive nature in the line of primates that eventually led to humankind, and suggesting that early humans in all probability shared at least some of those aggressive characteristics.[12]

Thus it seems that the Fertile Crescent had in all probability always been an essentially calm, reasonably tranquil land, and home to nomadic tribes of hunter-gatherer *Homo sapiens sapiens* for perhaps 70,000 years. Until about 10,000 B.C. the rhythmic flow of nomadic life throughout the region seems to have altered very little. The tribes moved, hunted, foraged, and occasionally clashed, but generally kept their distance. The sun rose and the sun set. Then suddenly — and over a period of no more than perhaps a few thousand years — a virtual explosion in human capacity changed everything.

Things never before seen or dreamed of suddenly mushroomed across the plains like spring flowers, many stunning and spectacular, others proving to be far more ominous. Here for the first time in human history small cities arose, surrounded by farm fields, rudimentary roads, and newly ordered cemeteries. Language blossomed and along with it primitive forms of both religion and philosophy. Nourished by years of good weather these small conclaves experienced a surplus of crops, in turn generating a meager form of personal property and wealth, and from that wealth the first symbolic form of exchange soon came into existence — money.[13]

This was not simply an accumulation of odd, unrelated events, nor was it an evolution of previously established technologies, traditions, or culture. In a very real sense this was the beginning of what we today call civilization. It was literally an explosion of human potential — a whole new world that sprang from nothing more seemingly substantial than the thin air of the Fertile Crescent — and out of the dark side of these stunning human developments something entirely new and frightening sallied forth across the virgin landscape. Somewhat remarkably, the advent of civilization appears to have arrived coincidentally with something truly uncivilized — war. "At the beginning of the New Stone Age, however, some 10,000 years ago, there occurred a 'revolution in weapons technology ... four staggeringly powerful new weapons make their appearance ... the bow, the sling, the dagger ... and the mace.' The last three were refinements of weapons already in existence: the mace derived from the club, the dagger from the spear point and the sling from the *bolas*, the last a pair of stones covered with leather and joined by a thong, thrown to entangle the legs of deer or bison which had been herded into a killing-place."[14] About this radical development in weapons Robert O'Connell poses an intriguing question, asking if there might in fact be a direct link between the development of weapons and humankind's physical and mental evolution.[15] Or, put somewhat differently, we might ask if there is a direct causal link between our own physical and mental evolution — the emergence of civilization, that is — and the emergence of war.

It is not known precisely when war first appeared, but it is known that war appeared coincidentally with early civilization and that what we call civilization began to coalesce into small towns and villages in the Fertile Crescent about 12,000 years ago. The best estimates suggest that war emerged a few thousand years thereafter. Robert O'Connell, for instance, in his study of humans and their weapons, states: "the best estimate is that war, true war, began somewhere between seven and nine thousand years ago — although it could have been much earlier — not as an aberration of the human psyche, but as the culmination of a revolutionary change in man's economic and social life."[16] And it was in the Fertile Crescent, in a landscape of remarkable abundance, that war first appeared. Never before on earth had war been conducted, yet it would haunt every year of humankind's existence for at least the next twelve millennia.

Since the beginning of the twentieth century, a long and involved debate has been ongoing over the nature of aggressiveness, violence, and war, so it may be wise at this point to define precisely what is meant here by the term *war*. "There is aggressiveness, arising from the competition of beings without which natural selection could not take place. There is violence, that form of aggressiveness which employs or effectively threatens the use of physical force. And there is war, that particular form of organized violence taking place between groups."[17] The present inquiry is concerned with the last category, specifically the root cause of predatory violence between organized groups of human beings. This book is about predatory human warfare. In that sense it should be understood that the simple archeological discovery of aggressiveness, weapons, or fighting among early *Homo sapiens sapiens* does not necessarily constitute war. As the American anthropologist Harry Turney-High theorized after studying numerous modes of tribal fighting all across the globe, "most of the societies which ethnographers preferred to study," he observed, "existed 'below the military horizon' and it was only when the sun of their future rose above it that they emerged into modernity.... All societies trapped at that level, he insisted, were bound to remain primitive until kingdom come."[18] Put differently, tribal societies might squabble and fight, but it would not be until they crossed what Turney-High termed the military horizon that they would develop the rudiments of a modern civilization. Once again, war and civilization seemed to have emerged coincidentally, virtually arm in arm, and that fact presents a troubling question. Why had the high arts of civilization evolved simultaneously with the low arts of war?

And it was on the warm plains of the Fertile Crescent some 12,000 years ago (give or take a few thousand years) where war first emerged, where the military horizon was first crossed. "The earliest organized military forces for which hard evidence is available are those of Mesopotamia, specifically the cluster of city-states ... known collectively as Sumer."[19] This fact has been established by the appearance of elaborate defensive systems around cities such as walls, towers, etc., and the sudden appearance in the archeological record of offensive weaponry such as the battle axe, mace, and lance — weapons clearly intended for combat, not hunting. Jericho, for instance, founded approximately 10,000 years ago, was surrounded by a massive stone wall and a stone tower some thirty feet in height.[20] This settlement constituted a stronghold, and as John Keegan points out, "A stronghold is a place not merely of safety from attack but also of active defence, a centre where the defenders are secure from surprise or superior numbers, and also a base from which they may sally forth to hold predators at bay and impose military control over the area in which their interests lie."[21] In short, a stronghold is a place not simply for defense, but a place for offense as well, where warriors can both enter and exit. It is a place of war.

The spears, arrows, and axes carried by these early warriors were all fashioned from stone and wood (bronze weapons will not appear until 3,000 B.C., iron until 1500 B.C.),[22] and on their belts the men carried slings to hurl rocks, accurate up to almost eighty yards — the very first form of artillery. These warriors were not yet professional soldiers — that too was still centuries away — but they were well practiced in the art of killing, and they were led by accomplished leaders.

From the hunting strategies of the nomadic bands it can be reasoned that the use of the natural features of the land to cloak movements and deployment were almost instinctively understood. Already the column, battle line, and envelopment — the fundamentals that would comprise battle for thousands of years to come — were understood, even if only primitively. In the years to come those ideas would be expanded upon substantially, the technologies of destruction improved exponentially; but the rudiments of war had been rapidly developed.[23] O'Connell points out, for instance, that most of the elements of battle were developed very early on and were at times then forgotten and subsequently rediscovered from time to time over the course of history.[24] Thus after a few million years of human and protohuman existence war finally arrived in the Fertile Crescent with all its organized and efficient horror. And the question that must be asked is ... why? And why here, and, more importantly, why *now*? How is it that a creature whose entire existence on earth never featured warfare suddenly exploded into full blown, organized violence? Just what can account for such an extraordinary transformation?

Make no mistake about the fury that was unleashed. Ghengis Kahn, Alexander, Caesar, Hitler, Napoleon, Stalin — to name but a very, few — were all direct descendants of the first line of battle that was drawn somewhere in the Fertile Crescent thousands of years ago. Moreover, when war appeared it did so with a sudden orgy of almost incomprehensible savagery. Indeed, so violent are the accounts of early warfare that it remains a moment in human evolution that is today almost impossible to comprehend. It is not simply that *Homo sapiens sapiens* abruptly mutated from an essentially peaceful animal into a violent one, but that this change arrived so suddenly and with such unprecedented barbarity.

Those are the cold facts that are so difficult to account for. Was this savage development the product of a genetic variant run amok? Or was the motive simply greed, a clash over resources, the result of a territorial imperative, as some suggest, or perhaps even some form of virulent pathology? This question is at the very heart of this inquiry, yet over the centuries the answer has proven elusive.

Unless we can come to grips with this one moment in time, understand just what brought about this shocking watershed event in the affairs of the human race, we will never really be able to understand war or, for that matter, ourselves as human beings. For we are all, in one way or another, the products of this critical juncture, and its ramifications still rattle our modern world like old bones hidden away in a hallway closet. Whether we like it or not, care to contemplate it or not, war is our heritage, and a robust heritage it has been. As the writer Robert Ardrey once put it, "Human war has been the most successful of all our cultural traditions."[25] *The* most successful. Why?

Interestingly, war did not just appear as a sort of singular, localized anomaly, only then to slowly develop, spread, and evolve like a malignant weed in the hands of other cultures and other minds. Rather it erupted across the globe, from Asia to the America's, full blown and monstrous, like some virulent seed waiting only to sprout from the beating heart of humankind whenever and wherever the conditions proved ripe. No race, continent, or culture proved immune to war's seductive lure.[26]

According to Trevor Dupuy, the first actual battle ever recorded was that of Megiddo, fought in Palestine in the year 1496 B.C. between local tribes in revolt against their Egyptian rulers.[27] Prior to that there is only conjecture. But if by 8,000 B.C. the city of Jericho required a wall ten feet wide to defend itself, it is not terribly difficult to imagine the forces already unleashed. The purpose of this inquiry is simply to try to grasp these ungovernable forces — this thing called war — to understand the pressures that seem to compel humankind to fight, all with the hope that this understanding alone might in some way prove key to war's ultimate undoing. As Konrad Lorenz once explained, "Wherever man has achieved the power of voluntarily guiding a natural phenomenon in a certain direction, he has owed it to his understanding of the chain of causes which formed it."[28]

The confusing issue over the ages for philosopher, historian, scientist, and theologian alike has been that throughout our brief history on earth humankind has proven to be *both* the most compassionate and artistic creature alive and simultaneously the most brutally destructive animal ever to set foot on the planet. Why both? If humankind were either one or the other we would, as a subspecies, I suspect, make much more sense. It is this glaring contradiction that baffles. But contradictions often contain the germ of surprising truths. As the philosopher Alfred North Whitehead once put it, "In formal logic, a contradiction is the signal of defeat; but in the evolution of real knowledge it marks the first step in progress towards a victory."[29]

So we will attempt to resolve the riddle of humankind's conflicted nature, to find our victory, and in so doing we will arrive at not only a deeper conception of war but also a far more sophisticated understanding of ourselves as human beings. For we are all the offspring of war, the children of this age-old calamity, and my fear is that we will never really achieve the lasting peace we so long for if our understanding of war remains superficial and, therefore, at best incomplete. This greater understanding in the long run, of course, cannot change the destructive events of our common past. But it just might change our common future.

Songs, chants, and hopeful injunctions, no matter how well intentioned, will not bring us peace. Understanding is critical. Only through a rigorous, clear-eyed grasp of war will any real, workable peace ever be achieved. Great ills like pneumonia or the plagues that swept away whole European populations, it may be recalled, were not cured by prayers or chants or pious admonitions, but rather by the application of scientific thought through medical prescriptions, no matter how rudimentary. So let us shine the bright light of human reason into this dark thing called war with the hope that it will ultimately lead us down the path toward peace.

One

Some Proposals on the Origin of War

The inquiry about to be undertaken is concerned with war. It is not, however, about war in general. Nor is it intended as a tactical overview of all those human conflicts that have raged since the dawn of time—if not an impossible task, then most certainly an exhausting one. Rather it is an investigation into the root cause of war, that primal motivation that seems to have caused men to hurl themselves at one another in bloody combat for thousands of years. Moreover, this inquiry is not based on theory, opinion, speculation or hopeful analysis, but rather the facts as they can be currently established. For centuries curious minds have speculated on the cause of war, and there has been no shortage in both the number and nature of the suspects proposed. All of these proposals offer something important, but they ultimately fail to identify the most critical factor. And as long as that factor remains unidentified, war will remain a mystery.

The theories proposed regarding the genesis of war do not fail because the minds that conceived them were weak. Far from it. Many of mankind's finest minds have toiled over the dilemma of human warfare, and their insights have been penetrating and substantial. The difference lies only in the fact that we today are more fortunate in terms of the historical, psychological, and scientific research that can be marshaled to the task. This availability of this is all that is really new, but in the end these insights will make all the difference. So this inquiry begins with a brief examination of some of the many ideas that have been historically proffered regarding the origins of war before it moves on to the most critical factor of all.

Of all species on earth, only two are known to war—humans and insects. Ants, as an example, lacking the developed brain capacity of human beings, fight as a result of instinctual commands. Ant wars—just as with their other social behaviors—are generated by the release of pheromones, chemical markers that communicate specific stimuli to the colony in general. Ants groom, hunt, even relocate entire colonies on the basis of nothing more than a chemical message.[1] Individual ants respond to the chemical stimuli just as hydrogen and oxygen bond under the right circumstances to form water. There is no thought process involved. Particular chemicals elicit particular behaviors. With humans, on the other hand, war is not quite that simple.

Since humans are animals — but animals that differ from other animals most particularly in terms of their conscious development — a number of sources have been suggested as to the possible root of human warfare. For instance, it has been thought war might be genetic in origin, a hardwired behavior from our long-ago mammalian past, and in that sense essentially the same as the ants' reaction to pheromones. Or it could be a learned activity, such as Jacob Bronowski's suggestion that war arose as a result of a pattern of theft on the part of hunter-gatherers to which agricultural communities then responded with organized violence. Or war might be psychological, and therefore a more recent expression of our developing brains.

It seems that the obvious flaw in the genetic possibility is that, if war were a hardwired behavior, the archeological record should have produced specific evidence of warfare ever since humans in their earliest form appeared on earth. It can be recalled, however, that the exact opposite appears to be the case. For the initial period on earth (and this, depending upon the subspecies chosen, can be calculated for a duration of anywhere from 130,000 to 2.5 million years) human beings lived in relative peace, and there appears to be no evidence whatsoever that earlier, protohuman species engaged in anything resembling warfare. The caves of StoneAge people are rife with elaborate paintings of the hunt, a few of conflict, but none of war. War came much later, and this simple fact seems to fly in the face of the genetic possibility.

The plausibility of war's being a strictly genetic trait discounts as well the very unique and fertile human capacity to think. War, of course, does not these days appear to be a particularly rational course of action. But thinking itself is by no means a strictly rational process, and thought, no matter how primitive, deficient, or even illogical, seems always to have been at the very heart of war. Thus "the idea of men gathering compulsively around engineering tables and churning out tank designs like bees building a nest is so ridiculous as to be rejected by even the most deterministic."[2]

The argument over aggression, violence, and warfare has been long, and contested and remains essentially unresolved. Why do human beings war? "Studies of individual and group behaviour take different directions," John Keegan notes. "They share, however, a common ground, to which debate eventually returns: is man violent by nature or is his potentiality for violence — about such potentiality there can be no dispute, if only because man can kick and bite — translated into use by the operation of material factors?

Those who hold to the latter view, loosely categorized as 'materialists,' believe that their perceptions demolish the naturalist position. The naturalists unite to oppose the materialists but divide sharply between themselves. There is a minority whose members insist that man is naturally violent; though most would not concede the analogy, theirs is the argument of Christian theologians who hold to the story of the Fall and the doctrine of original sin. The majority reject such a characterization. They regard violent behaviour either as an aberrant activity in flawed individuals or as a response to particular sorts of provocation or stimulation, the inference being that if such triggers to violence can be identified and palliated or eliminated, violence can be banished from human intercourse."[3] Is mankind naturally inclined to violence (and thus war) as a result of some undetermined biological factor, or is it the case that mankind tends to respond to particular situations or cues with violent behavior? That, fundamentally, remains the question, and it is actually a very simple question. Unfortunately, to date there have been no simple, or at least compelling, answers.

As Robert O'Connell points out, only with the arrival of agriculture and later politics would true war arise.[4] (That observation — that war appeared coincidentally with both politics and agriculture — will prove central to our inquiry, so it will temporarily be stored away and recalled later when appropriate.) Moreover, no species or subspecies in the long line of mammalian predecessors in the human evolutionary line — from chimpanzee to *Homo erectus* — displayed any overt inclination toward war — aggressiveness, of course, fighting certainly, but not war. So it seems that war is a *later* development in the long line of mammalian ascent, and thus it is highly improbable that genes alone can account for the root cause of human warfare. Score one for the materialists. But wait! Not so fast, some would argue.

In 1996 Harvard anthropologist Richard Wrangham suggested in his book *Demonic Males: Apes and the Origins of Human Violence* that human warfare was directly linked to certain warlike tendencies in chimpanzees. "Chimpanzeelike violence preceded and paved the way for human war," Wrangham insisted, "making modern humans the dazed survivors of a continuous, five-million-year habit of lethal aggression."[5] That assertion was then supported in a paper published in *Current Biology* by anthropologists John Mitani, David P. Watts, and Sylvia J. Amsler. They wrote regarding observations made of chimpanzee males attacking and killing neighboring chimpanzees from a bordering region while on what are described as boundary patrols, further the authors observed that these attacks led to an ultimate territorial expansion. "A causal link," the group claims, "between lethal intergroup aggression and territorial expansion can be made now that the Ngogo chimpanzees use the area once occupied by some of their victims."[6] These patrols are described as involving "considerable travel, but little feeding or socializing; patrollers are unusually silent and move in single file line, while attending to signs of other chimpanzees."[7] Upon occasion the patrols included extreme violence when the patrollers happened upon chimpanzees from another bordering region, and at times these violent encounters ended in the death of a bordering chimp who was attacked and beaten to death. Ultimately, then, the Ngogo group expanded into a portion of the territory previously occupied by the bordering group, and it is upon this violence and the demonstrated territorial expansion that warlike tendencies were attributed to the Ngogo chimpanzees.

Over a ten-year period, writes the Mitani group, "Seventeen of the 18 observed fatal attacks were made by coalitions of Ngogo males on patrol. Thirteen of the 21 cases of lethal intergroup aggression (61.9%) occurred northeast of the Ngogo Territory in a circumscribed region that corresponded to an area of heavy patrol activity. Four victims were adult males, while 9 others were immatures."[8] In that the Ngogo group of chimpanzees ultimately expanded their territory into the adjacent, northeast quadrant of the bordering chimpanzee population, the fatal attacks were subsequently interpreted as predatory and the outcome therefore intentional—intentional, at least, on some level. Mitani et al. continue: "Our observations indicate that chimpanzees at Ngogo have expanded their territory at the expense of a neighboring community.... These findings are consistent with the hypothesis that lethal intergroup aggression reduces the coalitionary strength of opponents living in adjacent groups, leading to territorial expansion by the aggressors."[9] The implication by both the Mitani group and Wrangham is that these fatal attacks represent warlike activities, and that, at least according to Wrangham, humanity has inherited the chimpanzee's predatory instinct for both violence and territorial expansion. Human warfare, at least according to this line of reasoning, owes it genesis to the chimpanzee.

Not everyone agrees. To begin with, it is important to understand that the attacks along the northeast border in question took place over a ten-year period and that only 13 of the lethal attacks took place along the "disputed" border. Of those 13, only 4 were against adult males, the remainder being against what the Mitani group calls immatures.[10] In light of these facts, it seems difficult to maintain the argument that these clashes—violent though they were—represented lethal intergroup aggression based on any predatory cognitive or genetic impulse—that the attacks were purposeful, that is, and that the purpose was territorial acquisition. The facts seem to indicate these fatal attacks were not intended or designed territorial incursions, but rather patrolling activities that simply stumbled headfirst into circumstances that were opportune. Indeed, the patrols do not rise to the level of what might be defined even as a raid, much less primitive martial activity. In fact, the ultimate acquisition of territory by the Ngogo chimps could not be attributed to predatory ambitions, and Mitani et al. admit as much: "One prominent hypothesis suggests that chimpanzees attack neighbors to expand their territories, and to gain access to more food. Two cases apparently support this hypothesis, but neither furnishes definitive evidence."[11]

John Horgan, director of the Center for Science Writing at Stevens Institute of Technology, writing for *Scientific American* disputes virtually all of the claims made by Wrangham and later supported in the Mitani paper. "Wrangham and other chimpanzee researchers," argues Horgan, "often present the rate of 'intercommunity killing' in terms of annual deaths per 100,000 population. Mitani, for example, estimates the mortality rate from coalitionary attacks in Kibale to be as high as '2,790 per 100,000 individuals per year.' But the researchers witnessed only 18 coalitionary killings. All told, since Jane Goodall began observing chimpanzees in Tanzania's Gombe National Park in 1960, researchers have directly observed 31 intergroup killings, of which 17 were infants.... All these violence statistics, according to an analysis published this year by the anthropologists Robert Sussman and Joshua Marshack of Washington University in Saint Louis (W.U.) are based on 215 total years of observations at nine different sites in Africa. In other words, researchers at a typical site directly observe one killing every seven years."[12]

The warrior chimp hypothesis also neglects the existence of the pygmy chimp, *Pan paniscus,* also known as the bonobo — an animal that is just as related to humans genetically as is the more common chimpanzee, *Pan troglodytes*— but which has never been observed in the act of coalitionary killing. "Frans de Wall, a primatologist at Emory University, suggested last year in the *Wall Street Journal* that bonobos may be 'more representative of our primate background' than are chimpanzees."[13] It is deWall's conclusion that the oldest known human ancestor, *Ardipithecus ramidus,* was a far less aggressive animal than chimpanzees or even the bonobo. This supports the premise that humans actually evolved from a line that was less violent than of the common ancestor that of the chimpanzee. Humans did not, in fact, inherit any tendency toward violence from the chimpanzee.

But the case against the warrior chimpanzees hardly ends there: "There is also no fossil or archeological evidence that our ancestors fought millions or even hundreds of thousands of years ago."[14] Indeed, anthropologist Erik Trinkaus states emphatically that there is no evidence whatsoever that "war is continuous back to the common ancestor with chimps."[15] Evidence of warfare would not come to litter the archeological record until much later, when "signs of group violence emerged in other regions in northern Africa, the Middle East, Europe, Asia, Australia and the Americas. The evidence consists of skeletons with crushed skulls, hack marks and projectile points embedded in them; rock art depicting battles with spears, clubs, and bows and arrows; and fortifications for protection against attacks. These relics indicate that warfare arose as humans began shifting from a 'nomadic existence to a sedentary one, commonly although not necessarily tied to agriculture.'"[16]

The Demonic Male theory appears, therefore, to be an interesting but highly speculative and unproven hypothesis as to the origin of human warfare. While certainly suggestive of further research, at this point its principal weakness is clearly that it flies in the face of both the fossil and archeological records, records that have reasonably well established a date for the genesis of human warfare as sometime around 10,000 BC. Moreover, the number of observed coalitionary killings appears to be exceedingly low (at best 31 observed attacks at numerous sites over a ten year period) to support any serious comparison to warfare. At the site where the Ngogo chimpanzees expanded their territory, there were only four attacks on mature bordering males over a ten-year period. What's more, the nature of the killings themselves appears to be the product of chance encounters, hardly characteristic of the strategic implications the anthropologists attempt to imply — that is, genetically driven violence with the predatory goal of territorial acquisition. Rivers flood their banks and forge new channels, after all, but no one deduces intent in the movement or predation in the result. At this point in time, the facts simply do not support the Demonic Male hypothesis.

The "territorial imperative" implied in the Demonic Male hypothesis was first brought to the attention of the general public by the writer Robert Ardrey in his book *African Genesis*. In that work Ardrey articulated his theory — much like Mitani and Wrangham — that humankind had in fact evolved from a line of killer apes, and that this was the underlying cause of war — that is, a genetic predisposition to defend a specific territory against any sort of threat. While humans may be somewhat territorial in their makeup, a territorial imperative would not make humans any different from any number of other species that are also territorial but do not war. Logically, therefore, the instinct to protect a given territory cannot be equated with warlike ambitions, per se. Wolves, for instance (like the Ngogo chimps), certainly mark and defend specific territories, but wolves do not muster into battalions and march off to war, and neither do the Ngogo chimpanzees.

What is more important, the territorial imperative is actually a reaction to aggression. Robert Ardrey puts it succinctly: "Human warfare comes about only when the defensive instinct of a determined proprietor is challenged by the predatory compulsions of an equally determined territorial neighbor."[17] In that sense, any territorial reaction would have to be considered a response to war or a warlike act, not a predisposition to aggressive warfare in the first place. Thus, to get to the root of war, what needs to be understood is not the territorial reaction, but rather the predatory compulsions that initiated the contest to begin with. Does the territorial imperative force mankind to seek territory as well as defend it? If that be the case, why, then, do not wolves — or any other territorial animal, for that matter — war?

In *African Genesis,* Robert Ardrey explains the nature of the territorial imperative. "The territorial drive, as one ancient, animal foundation for that form of human misconduct known as war, is so obvious as to demand small attention. When Sir Arthur Keith found himself too old for any active contribution to the Second World War, his broodings produced the marvelous volume, *Essays on Human Evolution,* and the conclusion: 'We have to recognize that the conditions that give rise to war — the separation of animals into social groups, the "right" of each group to its own area, and the evolution of an enmity complex to defend such areas — were on earth long before man made his appearance.' Such an observation of a human instinct probably more compulsive than sex throws into pale context the more wistful conclusions of the romantic fallacy: that wars are a product of munitions makers, or of struggles for markets, or of the class struggle; or that human hostility arises in unhappy family relationships, or in the metaphysical reaches of some organic death force."[18] But can the hidden, unconscious instinct to territory — either to take or to defend — truly account for war? What about wars where ideas are involved, or simply power as defined in strictly human terms, or a rush to defend the innocent? How can the territorial imperative explain these wars? It cannot.

Ardrey returns then to the seemingly confused and conflicted nature of humankind. How is it that war and civilization arrived simultaneously? How, exactly, can Hitler and Mozart, Alexander and Einstein be made sense of? As Robert Ardrey says:

> Had we been born of a fallen angel, then the contemporary predicament would lie as far beyond solution as it would lie beyond explanation. Our wars and our atrocities, our crimes and our quarrels, our tyrannies and our injustices could be ascribed to nothing other than singular human achievement.... But we were born of risen apes, not fallen angels, and the apes were armed killers besides. And so what shall we wonder at? Our murders and massacres and missiles, and our irreconcilable regiments? Or our treaties whatever they may be worth; our symphonies however seldom they may be played; our peaceful acres, however frequently they may be converted into battlefields; our dreams however rarely they may be accomplished. The miracle of man is not how far he has sunk but how magnificently he has risen. We are known among the stars by our poems, not our corpses.[19]

Whether we are known in fact among the stars for our poems and not our corpses is, of course, a matter of conjecture, but Ardrey makes an important point here nevertheless. As he puts it, mankind was born of "risen apes," and apes that were armed killers to boot. Yet these armed apes did not war in ages past, nor do they now. Whatever the Ngogo chimpanzees did or did not do, for instance, can be debated, but there can be little debate that their activity over a ten-year period did not rise to even a most basic and primitive form of warfare. Rather it was the more developed human being — capable also of poems, symphonies, art, and architecture — that invented the machine gun, the atomic bomb, and the blitzkrieg; it was the subspecies located much further along the evolutionary tree that discovered warfare, and not the more primitive ape. Can this be made sense of? Ardrey attempts to do so, but he comes up short: "I believe that civilization has come to mankind as neither accident nor ornament.... I regard it as anything but a coincidence that the rate of civilization's rise has corresponded so closely with man's ascendant capacity to kill. Civilization is a compensatory consequence of our killing imperative; the one could not exist without the other."[20] A compensatory consequence? Now, this is an interesting supposition, yet one that is critically flawed, not to mention illogical. We might simply ask, why? How, exactly, does civilization function as a compensatory consequence for mass murder? What is it about mass murder that makes us yearn for art — or vice versa, for that matter?

We come then to the idea that war erupted as a natural function of human learning, no different from the advances that have been made in, say, mathematics, engineering, or medicine. This is precisely what Jacob Bronowski had in mind when he suggested that war was probably nothing more than the natural response of town dwellers to the routine raids of hunter-gatherer tribes on their crops, herds, or surpluses. From the enormous distance afforded by historic perspective, this idea does seem somewhat reasonable. But it, too, is flawed.

To begin with, these were not times conducive to rapid learning or invention. Research indicates that advances in learning accumulated at a much slower rate 12,000 years ago than they do today. Indeed, today it seems almost impossible for professionals to keep up with the advances that occur in their own disciplines during the course of a single year, so rapidly is knowledge accumulating; but this was hardly the case thousands of years ago. A system of crude clay tokens, for instance, sprang up throughout the Fertile Crescent sometime around 10,000 BC. These tokens were marked with basic signs indicating types of crops, animals, etc., and were used by traders from Africa clear to the Caspian Sea. While used widely and constantly, this system — the precursor to written language, by the way — did not change appreciably for a period of almost 5,000 years.[21]

The first plows used by the Sumerians to work their fields were nothing more than oddly shaped tree branches dragged by men across their fields to cut rough furrows. It would take almost 6,000 more years for humans to develop a copper plow and then figure out how to hook that plow to an ox.[22] Likewise, in metallurgy it would require thousands of years to discover that copper could be extracted by means of smelting ore in an oven, and a thousand more before that same copper would be cast in molds.[23] The list goes on and on. This was not an era of rapid intellectual or technological advance. Yet, as Robert O'Connell has already pointed out, war did not slowly evolve over a long period of time but literally burst upon the scene, virtually full blown in scope and violence at its very inception. In light of the almost glacial rate of advance in these other cultural arts and practices — practices that were both widely and constantly in use and therefore should have been prone to productive tinkering — it seems almost impossible to imagine that war was a product of the normal human learning curve.

The argument has also been made that war came about simply as a function of various human populations exhausting their local resources, then clashing with other populations in an attempt to obtain greater advantage. While there is little doubt that various natural resources — most recently oil, for instance — have at times played into war scenarios with the rise of the modern nation state, the notion that war evolved as a result of primitive tribes clashing over natural resources flies directly in the face of the facts. Harvard biologist Edward O. Wilson dismisses the notion, pointing out that "war and civil strife have many causes, most not related directly to environmental stress."[24] Moreover, as has been demonstrated, some 12,000 years ago human populations were small due to the recent ice age, while natural resources had suddenly become remarkably abundant. Yet war still erupted during this period, a period when conditions were historically some of the *least* likely — at least according to the resource theory — to produce either movement or conflict. So it seems that aggression over natural resources could not have been the predatory trigger that instigated the initial march to war. And the argument goes on.

John Keegan points out that in 1986, for instance, the debate over war's genesis and humankind's nature came to a head at a meeting at Seville University where a statement was issued "modeled on the United Nations Educational, Scientific and Cultural Organisation's Statement on Race, condemning belief in man's violent nature in absolute terms. The Seveille Statement contains five articles, each beginning 'It is scientifically incorrect,' to which affirmation is expected."[25] Unfortunately, political manifestos, no matter how well intentioned, cannot enforce a view of humankind for which science and reason have thus far consistently failed to supply firm evidence — either for or against, and attempts to legislate reality, in that their impulse is to restrict inquiry and honest debate rather than encourage it, often do far more harm than good. Humankind's true nature will not be determined by scientific fiat but rather by scientific fact.

But one thing is certain. Any explanation of war must of necessity be capable of explaining three critical and seemingly conflicting facts: (1) Why did war and civilization arise coincidentally? (2) Why is war an attribute of the more developed subspecies known as *Homo sapiens sapiens*, rather than its less developed ancestors? And finally (3) Why has the motive for war evolved over the centuries? (This last is a fact that will be established in later chapters.)

Summing up his exhaustive study of war published in 1967, Stanton Coblentz stated that "in order to know what war really is, and how and why it acts as it does, one must look beneath the physical mask to the psychological truth...."[26] Our survey of three thousand years of war-making has shown us that war has been largely if not entirely the product of psychological urges."[27]

What Coblentz discovered was that the fundamental motivation for war was psychological, not learned, territorial, resource driven, or genetic. That psychological motivation then plays itself out in a host of secondary impulses, from greed and economic objectives to racial and territorial imperatives, various great causes, and on and on. But the initial motivation to war appears to be psychological, and that is precisely what we will discover as we continue our inquiry. And since humans are the most psychologically sophisticated of all mammals, this also makes sense as to why war emerged later rather than earlier in terms of human evolution. So if we can identify this psychological function — this predatory compulsion, as Ardrey refers to it — then we will be able to identify the root cause of war.

Thomas Mann once remarked that "War is only a cowardly escape from the problems of peace."[28] If this is true — and I believe it can ultimately be demonstrated to be so — then the psychological response Coblentz refers to would have to be considered in essence cowardly or at least highly inappropriate. In that sense, then, war would seem to be a poor or weak (or cowardly) reaction to a strong psychological impulse.

To understand just how that might have happened some 12,000 years ago, we first need to understand exactly what sort of peace reigned for that long period of humankind's initial existence on Earth. What was it like before that first long column of armed men marched forth? Who were these hunter-gatherer *Homo sapiens sapiens* who lived in relative harmony for so long? What was the nature of the peace they enjoyed, how was it maintained, and finally, of course, what was the psychological impulse that caused it to finally collapse? In short, it seems impossible to understand the end of the tale if we cannot grasp the beginning.

As pointed out in the prologue, based on the fossil record, from a physical or anatomical perspective, early *Homo sapiens sapiens* appear to have been remarkably modern creatures. They looked, walked, and, perhaps, were beginning to talk very much like modern human beings sometime between 15,000 and 10,000 BC. Their technology and culture, as we all know, lagged far behind our own, and from what can be discerned from their recovered artifacts, so did their psychological makeup. How can this be established?

In the second half of the twentieth century an enormous amount of information regarding human psychological development was discovered. Naturally, a significant portion of that information can be applied to early *Homo sapiens sapiens*, just as it can be applied to today's modern human development. Much of this information will be covered in detail in later chapters, but a fundamental rundown will be provided in the next chapter regarding those stages that have directly affected the genesis and continued development of war.

Interestingly, while many books have been written regarding war and aggression in general, very few have included in their presentations this critical information. Indeed, during the 20th century a small cottage industry concerning war blossomed that used a range of new methodologies and analyzed war from almost every conceivable angle. Content analysis (a method used to measure variables in written and verbal communication), comparative case studies, international simulations, computer simulations, social science extrapolations, even mathematical models were utilized, and all with at least some interesting results.[29] Unfortunately, all of these methods employed certain assumptions, and if those assumptions were even partially incorrect then the results were naturally skewed. Moreover, virtually all of these studies focused almost exclusively on modern warfare and international strategic objectives, and few if any included data of a psychological nature. By and large, this was due to the fact that these important psychological studies were not available until relatively recently. But what these studies tell us about war will prove critically important.

Here is another interesting point: Almost all previous inquiries have treated war as a sort of singular, boilerplate event that, evolved in technology and perhaps ferocity, but like molten iron or gold, each and every last one was poured from the same bubbling cauldron. However, what we will discover is that this is not the case at all. What will be demonstrated is that as broad psychological outlooks alter for given groups of human beings, the nature and intentionality of war also alter. War has not been evolving only from a technological, strategic, and tactical perspective. It has also been evolving in accordance with humankind's developmental framework, a framework that only recently has come to be understood in a far more sophisticated manner than in previous centuries. In a very real sense, that is the encouraging news, because humanity is developing and that fact provides room for hope. While this is not a work about human development per se, the argument here will be made that war itself, as well as war's evolution over the centuries, has been psychologically driven. Thus, to grasp the nuts and bolts of the argument it will be necessary to understand at least the fundamentals of the early stages of human development, those stages that have created war and that have continued to affect it over time. The next chapter, therefore, will be dedicated to a brief explanation of these initial stages so that the reader might sensibly apply them to the remainder of the inquiry as it unfolds.

Two

The Developmental Path

As we mature as human beings our psychological makeup changes considerably. Those changing psychological makeups cannot help but have a profound effect on how we conceive of ourselves and our relationships with one another. This is true for us as individuals, and it is equally true for groups who share the same psychological outlook.

There is general consensus among today's most prominent developmental psychologists and social scientists — Erich Neumann, Ken Wilber, Jean Gebser, Jurgen Habermas, Don Beck, to name but a few — that human consciousness (that apprehension of reality that you, I, and most humans now share) was not fully present by and large in the first human beings, but rather slowly emerged over the long span of human existence and that this emergence came about in roughly four distinct stages or epochs. Naturally, these stages began with the most limited, or least robust, form of consciousness, and expanded exponentially into the forms of consciousness currently known in this, the most modern epoch.[1] Carter Phipps puts it this way: "These stages are psychological and cultural levels of development; they are levels of consciousness that individuals pass through in their personal evolution and that societies pass through in their cultural evolution."[2]

Research has demonstrated that these stages are universal. They have applied to all races, regions, and cultures on earth, and the cultural stages very closely resemble the developmental stages each individual human being passes through as they grow to maturity.[3] Many theorists have used different names for these stages. One of the original thinkers in this area — a name many readers are probably familiar with — was the Swiss developmental psychologist Jean Piaget. Through the exhaustive study of young children, Piaget was able to carefully detail four distinct stages of development. As Philip Rice puts it, "Piaget showed that from birth onward, intellectual competencies undergo continuous development until they obtain their adult forms. And, as his later work emphasized, development never ends."[4] Thus there is every reason to believe that these stages will continue to unfold as human beings continue to develop. As Jean Piaget laid down the groundwork for these stages, and as the formulations in this inquiry regarding warfare are predicated upon the changes that occur as humans move from one stage to another, it is essential to understand at least the basic outline of the formative stages as conceived by Piaget so that the reader can then critically evaluate information supplied in later chapters.

"Piaget called the first stage of human development *sensorimotor*. This period lasts from birth to approximately the age of two. The newborn child moves from a self-centered, body-centered world to an object-centered world. Thinking, if any, occurs in the form of a stimulus-response connection with the physical world. The latter aspect of this stage is marked by a transition to symbolic play, imitation, and the representation of objects. In short, the newborn child's world is entirely self, or body based. The world is comprehended — if at all — as merely an extension of the self."[5]

Today Ken Wilber is considered by many to be the world's leading expert on human development, and this is what Wilber has to say about *sensorimotor* (a stage that we will from this point on refer to for simplicity's sake as level one): "If all goes relatively well, then the infant *transcends* this archaic fusion state and emerges or hatches as a grounded physical self.... The *sensorimotor period* (0–2 years) is thus predominantly concerned with differentiating the *physical* self from the *physical* environment."[6] From a cultural perspective, this level one portion of the human epoch was obviously the earliest and thus the least developed:

> Family or clan relationships no doubt existed, and in the later stages of this epoch, tribes began to flourish. Because the distinguishing cognitive feature of this period is a body-centered or completely self-centered understanding of reality, any social intercourse would have been limited and generally inarticulate. Hand gestures and the very beginnings of language would dominate this period. Any art, theology, or language would have been extremely primitive and body-based.[7]

For the child who successfully negotiates this first stage of development (and some do not due to genetic problems, severe illness, or repeated trauma) the child then begins to enter the next stage, which from now on we will refer to simply as level two.

> Piaget called the second period of enhanced cognition *pre-operational*. The child is involved with pre-operational thinking from the age of two to approximately seven. The pre-operational stage is dominated by the acquisition of language and the utilization of symbols. Thinking during this period remains egocentric and self-centered. The child has a difficult time understanding the reason that everyone does not see all the facts the way he or she does. The child also has great difficulty focusing on all the varied aspects of any given situation.[8]

It should be stressed that the shift from one stage of development to another is not rapid, but generally requires a period of transition, aspects of the old stage still dominating the new. According to Ken Wilber, "At the end of the sensorimotor period, the physical self and physical other are clearly differentiated, but as the *mind begins to emerge* with preop [preoperational], *the mental images and symbols themselves are initially fused and confused with the external world*, leading to what Piaget calls 'adherences,' which children themselves will eventually reject as being inadequate and misleading."[9] By adherences, Piaget was referring to aspects of the sensorimotor stage that continue to dominate the early stages of preoperational thinking. Piaget explains: "There are, to begin with, during a very early stage, feelings of participation accompanied sometimes by magical beliefs; the sun and the moon follow us, and if we walk, it is enough to make them move along; things notice us and obey us, like the wind, the clouds, the night, etc."[10]

This portion of the human epoch displayed a marked development over its predecessor.

> This stage is dominated by tribal communities with strong ties of kinship. The individual is merely an extension of the kinship bond. The clan or tribe, in turn, sees itself as a natural extension of a nature that functions through magical interactions. The individual, the tribe, or nature is not yet clearly differentiated, thus, physical objects may take on the aspect of living entities, and natural events, such as storms, floods, or droughts, are at times confused with mental projections.[11]

Another important development during this period is the shift from magic causality to mythic. "Maybe I can't order the world around," as Wilber puts it, "but Daddy (or God or the volcano spirit) can.... And thus come crashing a hundred gods and goddesses, all capable of doing what I can no longer do: miraculously alter the patterns of nature in order to cater to my wants."[12] And what is important for our inquiry, war does not exist at levels one or two. War, as we shall see, is a later development in the evolution of human consciousness.

The next stage of development Piaget called *concrete-operational*, and this stage rather naturally entails a significant expansion of cognitive abilities. Myth dominates thinking during this phase of growth, while thinking itself is concrete and absolute. This, then, is an epoch of black and white, with few shades of grey, and this stage we will refer to as level three. The myths of the era were not considered fantasy or allegory but truth absolute. "If the sensorimotor and preoperationsl world is egocentric and geocentric, the concrete operational is sociocentric (centered not so much on a bodily identity as on a *role* identity)."[13]

This was an epoch of empires capable of banding together the local tribes into communities with an allegiance to central authority. That central authority

> had both validity and access via the gods. Myths grew to explain the source of this legitimacy and, over time, the distinction between the head of the state and the gods dissolved entirely — the head of the state *was* a god. If the head of state was a god, of course, then the state was indeed infallible because gods do not err. Thus did the great kingdoms arise that were to rule the planet for centuries, vestiges of which remain today.[14]

This is the level in which war appears, and when it does so it appears with a suddenness and brutality that are breathtaking.

The last of Piaget's four initial stages of growth is named *formal operational*. For most individuals, formal operational is entered during the early teenage years up to early adulthood. Formal operational presents an entirely new, robust form of thinking; and due to the changes inherent in the process, this development, for many teenagers, is a time of upheaval and anxiety. Thinking changes gradually from the absolute thought patterns of concrete operational to rationality — to logic. The ego here first arises, and thus young adults generally have a period where they must "find" themselves; that is, they establish their true identity or ego. It is a stage of great possibilities, but also of great changes. Wilber puts it this way: "First and foremost, formal operational awareness brings with it a new world of feelings, of dreams, of wild passions and idealistic strivings."[15] As he explains it, armed with this new rationality the individual for the first time is able to take the place of the "other" and to envision just what might be.

This is the reason that idealism arises generally during formal operational, for the individual can grasp all potential possibilities and compare them to the mundane, insufficient present:

> In cultural terms, this translates into tolerance, while, in social terms, into the first democratic notions. Thus, this period of human history is marked by upheaval. Concrete, literal-minded empires do not take well to the impertinence of scientific, rational thought. They are not about to toss aside their birthright to the very rabble whose own sole legitimate purpose is to serve *them*. Revolution after bloody revolution ensues. The world is swept by the battle cry of liberty, of demands for God-given rights and freedoms. The results are ugly and bloody, as we still see in our world today.[16]

Science, at last free to investigate the physical world, blossoms, the scientific method is discovered, and this new science of logic, testing, and discovery soon clashes headlong with level three religions based entirely upon dogmatic myth. "Republics arise, nations born with seminal egalitarian notions as their guiding principles. Over time, these notions will be expanded, becoming far more inclusive, fair, and compassionate. Monarchies either collapse or become irrelevant."[17]

These, then, in very rough outline are the first four stages of growth. The two that will dominate our inquiry are levels three and four. It is impossible to truly understand war without a grasp of these levels. Indeed, it is impossible to understand the long, often sanguine history of *Homo sapiens sapiens* without understanding the dynamic changes that affect human beings as they move through these epochs. The patterns of thought—the worldviews—that dominate each of these epochs shape the art, religion, philosophy, culture and social organization of each and, as we shall see, they also shape the intention that drive humans to war. Armed with an appreciation of the basics of developmental psychology, we see that history begins to collapse into recognizable form, and the world begins to make sense. For simplicity's sake, we are going to note the fact that historically there have been roughly four sequential stages, each of which has been built upon the preceding stages. As outlined above, we are going to call these levels one, two, three and four.

In a groundbreaking study published in 1995, the psychiatrist Clifford Anderson confirmed Piaget's findings while at the same time adding some interesting and important findings of his own: "In everyday life an incremental expansion of range may go unnoticed. But in the overall process of psychological maturation, there are four major identifiable stages in range expansion."[18] As noted, the four major stages Anderson's study revealed were the same, basically, as those uncovered by Jean Piaget. Each new stage ushers in a new, far more robust cognitive format than the stage before it; while growing up most of us have passed through these stages on our way to adulthood.

Just as in Piaget's studies, the first stage Anderson notes—that of the newborn infant—is a very limited form of consciousness. Of the three following stages Anderson suggests that in a general way "concrete thinking indicates childhood; abstract thinking indicates adolescence; and relativistic thinking best indicates youth."[19] He continues: "Finally, it should be reemphasized that these psychological advances were not accomplished in the classical evolutionary sense. Rather, they reflect movement down a preexisting path—a template every bit as inherent as the physical one that creates the brain, the heart, and the muscles in the arm. Moreover, it is a path whose potential has yet to be fully realized."[20]

Research has shown that a person's psychological growth is a combination of their individual traits and the maturational conditions in which they find themselves.[21] As Anderson indicated, most people don't even notice these stages as they pass through them, although the transitions can be bumpy at times. Generally speaking, advanced growth requires optimum conditions, and since the conditions during the late Stone Age were far from optimum—in terms of food, shelter, personal safety, education, medical care, and the like — it is entirely reasonable to assume that most people did not mature much beyond levels one or two.

In that sense, humankind, just like the individual human being, appears to have had an extended period of psychological infancy. Anderson explains:

> Placed in historical perspective, this model suggests that before the twentieth century most adults in the general population remained psychologically arrested in childhood. Although they continued to grow up physically, their thinking remained concrete or literal. The stages of life in which higher thought forms are constructed simply never occurred — an inference that partially explains the harsh, rigid world our ancestors constructed for themselves.[22]

This fact is absolutely critical for understanding the origins of human warfare and is often overlooked during historical, philosophical, and spiritual analyses of human history.

The psychological stage that dominated the hunter-gatherer epoch — and that prevailed for perhaps as long as one hundred thousand years prior to the coming of war — was level two, the second stage of growth listed by Clifford Anderson and what Piaget termed *preoperational*. Here again is how Ken Wilber describes this stage:

> At this early stage (level 2) the logical, verbal, and conceptual mind is not yet developed (level 3/4). The mental capacities, such as they are, are simple and crude by any standards, consisting basically of primary process or magico-imagery, paleosymbols, and protolinguistic structures. Since the mind is not yet developed, it does not have the capacity to differentiate itself from the body, and thus the self likewise is embedded in and undifferentiated from the body.[23]

What Wilber is saying is that at level two (hunter-gatherer) the self was not yet a mental self — such as most healthy individuals experience today — but rather a self where the emerging mental capacities were still experienced as part and parcel of the body; that is, they were not yet differentiated from the body. This is an extremely important point for understanding all human development. We literally experience a different form of "self" at each level of development we engage, and in this inquiry we will return to this point again and again.

The level two self was, at least by more modern standards, a very primitive and somewhat confused self. The individual life was experienced not as the clearheaded mental ego most individuals experience in their lives today but rather as a mélange of body and mind, mind and environment, environment and body. Wilber describes it succinctly: "The angel and the animal, the man and the serpent, were one."[24] According to some Native American traditions, for instance, the mother of a newborn baby would often name the child after the objects encountered soon after birth — thus names like Morning Cloud, or Small Calf, or Sitting Bull. The names were not meant as unique descriptions or personal appellations in the modern sense. Rather they were continuations of the greater environment, a fusion of self and nature. This fusion of self and nature is captured by the Native American writer N. Scott Momaday:

> For the Indian conceives of himself in terms of the land. His imagination of himself is also and at once an imagination of the physical world from which he proceeds and to which he returns in the journey of his life. The landscape is his natural element; it is the only dimension in which his life is possible. The notion that he is independent of the earth, that he can be severed from it and remain whole, does not occur to him; such a notion is false and would therefore be unworthy of him.[25]

This inability to differentiate at level two led directly to a world dominated by what were interpreted as magical interactions. As Ken Wilber says, "At this early stage, then, although the self is distinguished from the naturic environment, it remains magically intermingled with it. The cognitive processes at this stage thus confuse not only subject and object, but whole and part."[26]

What was the world of the hunter-gatherer like? It was a world of shamans, of sorcerers, of rain dances and hunting rituals. It was a world of wind, water, and thunder gods, all acting magically to move and alter the world. It was a world where the hunter shared a natural, magical bond with the prey, and where the mountains had thoughts and the wind spoke its wisdom. It was a world where the image was confused with the actuality, so that a shaman wearing the head of a wolf, the teeth of a lion, and the feathers of a hawk was thought to actually possess the powers of all three — to magically *be* all three.

It was a world where similar things were considered identical (all members of a particular tribe were considered untrustworthy if one of the tribe was known to be untrustworthy), and where proximity was often fused with identity, thus any part of a thing carried the essence of the whole.[27] In just this manner the scalp of a rival tribe member was thought to contain the power of the individual from whom it was taken; the good fortune of the hare could be carried in the form of a rabbit's foot; and the hunting proficiency of one distinguished hunter could be assumed by another simply by stealing the former's lance, his bow, or his club. To make a doll-like image of a rival and then stick a pin or knife into the doll was thought to produce ill effects in the actual rival because, according to the thought process of the era, similar things were considered identical.

While we modern, cosmopolitan people tend to go about our lives in a generally more sophisticated fashion, we are really not so advanced or so far beyond the hunter-gatherer stage as we would like to think. Consider the pitcher on a winning streak who refuses to wash his socks for fear of failure; voodoo, all forms of irrational prejudice (similar things are considered identical), the good luck charms and the religious icons we casually wear or carry. All are Paleolithic forms of thought left over from our long-ago past that are still very much alive and well even today.

This confusion, this Paleolithic fusion of self and environment, is very evident in the art forms of the Stone Age period. Take, as an example, the simple totem pole. The carefully fashioned pole represents a sort of composite of the individual human and nature at large, the individual directly linked and literally composed of the animals he hunts. "Man's original fusion with the world, with its landscape and its fauna, has its best known anthropological expression in totemism, which regards a certain animal as an ancestor, a friend, or some kind of powerful or providential being."[28] Indeed, this relationship is often confused to the point of identification. In that sense totemism might be considered one of the world's first and most primitive religions, not a religion of doctrine or dogma, but rather a religion of powerful unspoken — indeed, subconscious — beliefs.

Likewise, the performance and dress of the ancient shaman appears clearly to be an expression of the same environmental fusion. The shaman represents not only the human presence but also the presence of a host of other birds and animals, all combined in a wild mélange. His is the dance of a vibrant and compound existence. Moreover, as Wilber points out, this was not a symbolic gesture, but rather a literal expression of Paleolithic self-comprehension. This was not a dress-up party, like our modern Halloween. This is how Stone Age humans actually envisioned themselves, literally empowered by the animals they aped.

This confusion, so obvious in both the totem and the shaman's dress, demonstrates clearly the nature of the worldview generally held by Stone Age tribes. While this worldview was substantially more robust than the one that preceded it (level one), it was, generally speaking, far less sophisticated than ours today (this, by the way, does not mean that all individuals during this period developed only to level two and no further, but only that this level was the general average for the epoch), and this tells us something essential about the peace that reigned in human affairs during these Stone Age times.

This peace, the long epoch of tranquility that preceded the era of organized warfare, was not in fact the product of natural decency or any process of rational self-interest. This period of harmony did not come about because thoughtful, well-intentioned hunter-gatherers sat down together and hammered out a sensible, Stone Age compact of peaceful coexistence. The self at this stage was simply not yet sufficiently developed or differentiated to be

The totem pole, a composite of human and animal features and a semi-religious icon, demonstrates the fusion of early hunter-gatherer societies with nature (NOAA, Family of Commander Leonard C. Johnson, USC & GS).

North American Native medicine man, or shaman, exorcising evil spirits; a melding of man and animal, spirit and reality, the shaman demonstrates the undifferentiated thought patterns of early hunter-gatherer societies (Library of Congress).

capable of grasping the issues inherent in its own self-interest on that sort of scale, much less consciously implementing such an agreement.

The peace that prevailed for millennium upon millennium prior to the arrival of war was the peace of slumber, not the peace of conscious, rational choice. It was the peace of human infancy, exactly the same as that enjoyed by the individual human infant, who once fully grown may still develop into someone far more violent. In that sense it was the same sort of peace enjoyed for eons by the deer and the fox, the elk and the bear. It was the peace of not knowing, not the peace of knowing better, and in that sense it seems clear now that humankind did not war during this epoch because our ancestors had wisely eschewed violence in favor of brotherly love, but simply because — as shall be soon demonstrated — they were not yet capable of this thing we call war. That may at first seem a strange assertion, but the fact of the matter is that war is a function of greater, not lesser, development in human beings. And to begin to understand the source of war, it is absolutely essential that we grasp that one critical fact.

The long period of human slumber would come to an end, and that awakening would prove both rapid and extraordinarily violent. Soon a creature that had been utterly incapable of large-scale, organized violence would virtually swim in a sea of its own blood. War would arrive suddenly, and it would arrive with thunder and savagery. Listen, for instance, to the boasts of a few early Assyrian leaders who ruled in the Fertile Crescent. First is Tiglath Pileser, from the twelfth century B.C.:

> The bodies of their warriors like the storm-god I hurled to earth, their blood in the ravines and on the heights of the mountains I made to flow down. Their heads I cut off, by the side of their cities like grain heaps I piled up. Their spoil, their property, their possessions, to an unnumbered quantity I brought out.[29]

Then comes Ashurnasir-pal:

> I cut off their heads. I burned them with fire, a pile of living men and of heads over against the city gate I set up. Men I impaled on stakes. The city I destroyed, devastated. I turned it into mounds and ruin heaps, the young men and maidens in the fire I burned.[30]

And there is King Sennacherib:

> Like the oncoming of a storm I broke loose and overwhelmed it like a hurricane.... With their corpses I filled the city squares.... The city and its houses, from its foundations to its top, I destroyed, I devastated, I burned with fire.[31]

Or read God's own instructions to the tribe of Israel as recorded in the book of Deuteronomy from the Old Testament:

But if it makes no peace with you, but makes war against you, then you shall besiege it; and when the Lord your God gives it into your hand you shall put all its males to the sword, but the women and the little ones, and the cattle, and everything else in the city, all its spoils, you shall take as booty for yourselves; and you shall enjoy the spoil of your enemies, which the Lord your God has given you.[32]

From these few passages it becomes painfully apparent that virtually overnight something radical had changed an essentially docile wanderer into a brutally savage and organized killer. It was something so critical that it was part and parcel of what Ken Wilber would call the most important transformation in the history of the world.[33] What could that transformation have been?

Three

Civilization's Dawn

Sometime around 10,000 B.C. civilization arrived in the Fertile Crescent like a clap of thunder, and right along with it came violent, organized warfare for the very first time. Since both politics and agriculture represent essential components of what we today call civilization, it is time now to pull out Robert O'Connell's observation from the preceding chapters — that war arrived concurrently with these other two — for further consideration.

After long and studied reflection on this curious turn of events many thinkers, from Jean-Jacques Rousseau on down, came to the conclusion that *civilization* itself must have been the corrupting agent that ushered in not only war but also a host of pathological conditions that ultimately robbed humanity of its pre-civilized bliss. From this line of reasoning emerged the notion of the "noble savage," that pure and regal original human being that had been compromised, corrupted, and ultimately enslaved by the agents of civilization. Surely the most famous statement of this line of reasoning is the first sentence of Rousseau's *The Social Contract*, where he declares that "man is born free; however he is everywhere in chains."[1] Two things are here clearly implied. The first is that mankind's original or prior state was one of natural freedom, hence one that was naturally proper or just. Second, this perspective assumes that mankind has in some manner been removed from this prior free, proper, and just state and has as a result been enslaved by the vagaries — economic, social, legal, etc., mechanisms — of civilization. From this perspective, then, it became easy to conceive of the less conscious hunter-gatherer era as in fact far superior to the more conscious, but also far more troubled, era ushered in by civilization. Thus a return to those earlier days represented the true path to human peace, prosperity, justice, and freedom. According to this worldview the course forward was, well — backward!

It is also easy to see now that it was this thought process that led rather naturally to the notion of a primordial Eden, that place where things had been more or less perfect long before civilization arrived and from which humanity had suffered a rough and inglorious fall. The problem facing mankind then — at least from this perspective — was not one of growth or looking forward to a better day, but rather racing backward, fleeing civilization, regressing actually, going back to that ancient Eden where humanity might recapture the goodness that had been lost so long ago — our freedom and natural dignity. This idea, both consciously and subconsciously, haunts much Western thought, from the Age of Enlightenment clear through today's postmodernism, and is something that will be critically addressed in the next chapter. But for the time being the question that must be asked if this inquiry is to pursue a sensible course at all is this. If civilization is the culprit that ushered in war, where exactly did this thing we call civilization come from? What caused *civilization*?

Webster defines civilization as "an advanced state of human society, in which a high level of culture, science, and government has been reached."[2] The problem is, for almost one hundred thousand years *Homo sapiens sapiens* had been living as hunter-gatherers, roaming the countryside in search of game, communing with the spirit-gods, and worshiping the totem. There is absolutely nothing in that formula that even suggests the blossoming of civilization. Yet suddenly men and women were living in small towns numbering up to a few thousand people, engaging in philosophy, commerce, politics, and war. How did that happen? There is a tendency to think that civilization was simply a natural step in human progress, but there is no line of reasoning or set of facts that actually supports that assumption. The sudden appearance of small cities is actually a rather curious and unexplained phenomenon. Suddenly people had begun to congregate, forming in small towns and cities, acting in ways human beings had never acted before. What caused all of this?

Some people have argued that war forced the hunter-gatherer societies into cities for their own protection. But it will be recalled that these nomadic tribes were incapable of war, so that argument really makes no sense — it is putting the cart before the horse. Others have argued that cities evolved naturally, and that humans simply settled down because it was the next logical step in a long line of natural ascension, caves being supplanted by grass huts, huts giving way to wood and animal skin shelters, and those shelters eventually evolving into small cities.

But we have already seen that the learning curve during these times was very slow, and we have also seen that the old hunter-gatherer society had given no indication whatsoever of change. Ken Wilber, as an example, objects strenuously to this line of reasoning: "That had *never*, at *any* time, happened before! The only analogy I can think of is to imagine dozens of hunting packs of wolves gathering together and settling down in a town of 200 wolves, and striking up social intercourse."[3] Wilber is entirely correct, of course, and it is only because we today almost intuitively grasp the evolutionary nature of the human experience that this thought process seems to make sense to us. The fact is, however, that the formation of permanent towns was a momentous and seemingly inexplicable alteration in the affairs of human beings. What caused it? Agriculture seems to be at the heart of the story; but farming is itself a civilized activity, and the hunter-gatherers showed no knowledge of or inclination toward farming. Fortunately, we can piece together a reasonably sensible story line as to what transpired, and that story goes something like the following discussion.

As the ice age glaciers slowly receded, the area of the Fertile Crescent blossomed as never before. Vast fields of wild cereals — wheat and barley — spread over literally thousands upon thousands of acres.[4] It was an astonishing bounty, food for thousands upon thousands of herd animals. With freshly melted glacial water tumbling through the nearby creeks, it was a time of unprecedented growth and prosperity.

The nomadic hunters naturally followed the herds and soon stumbled upon this enormous expanse of food and prey, a bonanza that at the time, I'm sure, probably seemed almost beyond comprehension. It would soon have been discovered, as well, that no amount of additional wandering would serve to improve their lot, so the fortunate tribes that originally discovered this windfall would soon have stopped and erected their usual shelters. In time they might have shifted here and there as more advantageous campsites were discovered, places with more firewood or better natural springs, for instance, and there they would have, for the time being at least, settled in.

As numerous historians point out, the wild wheat that grew in the region — einkorn and emmer — was extremely rich in protein, each acre yielding almost 700 pounds of consumable grain a year. A single family could strip enough during the three-week harvest period to fill their needs for an entire year, with a surplus to boot.[5] The wheat had to be threshed, sifted, and then ground to prepare it for cooking, but with a little work the whole tribe now had a steady and reliable source of highly nutritious food at their fingertips.

With more than ample supplies of water, grain, and prey in the area, there would have been no need to move on, so the nomads tentatively settled down. During the first few harvest seasons the women worked the fields while the men continued to hunt. As a result, for the first time since humans walked the earth, children no longer had to go out on the daily or weekly trek but could remain behind in a more or less permanent home with the elders of the tribe while their parents were off during the day. And that small, unintended lifestyle alteration would be the critical kernel that would blossom into what Wilber and other psychologists have called the greatest single change in the long history of humankind.

For hundreds of thousands of years humans had wandered the earth in search of food, dragging their children and elderly behind them — a harsh and difficult life by any standard, and a lifestyle hardly conducive to optimum cognitive and psychological growth for the young. Then suddenly all that changed, and radically for the better. As pointed out in the last chapter, psychological growth is dependent upon both genes and conditions (nature and nurture), and once the maturational conditions improved radically, it would not have been long — perhaps no more than a few generations — before the children cared for in this new manner broke through to level three psychological growth. With that development a whole new world suddenly dawned.

This new level of consciousness was markedly more robust than its predecessor, although in terms of analytical thinking it would not go beyond the concrete or literal, just as Anderson pointed out in chapter two. But level three consciousness would perceive the world on a much greater scale and with much greater understanding. As a result a whole sea change in human conditions would flower almost overnight. *Those* changes would cause what we today call civilization.

The initial discovery — and one that would initiate a vast feedback loop — was agriculture. It has already been noted that the processing of grains entailed tossing and sifting. During the course of all this work some of the seeds would naturally have dropped around the settlement or would have been tossed aside once the day's work was done. Over the course of months and years many of these seeds would have sprouted naturally, not where the old fields of wild grain were located but on the cleared ground along the edges of the encampment where they had never sprouted before. This simple but remarkable fact would not have been lost upon the emerging, level three consciousness, and not long after that the science of agriculture surely blossomed for the first time in the human mind.

From this new science of agriculture — even in its original, primitive form — would come an abundance of food, no longer snatched from a fickle and sometimes uncooperative nature, but produced instead by the freshly energized mind of humankind. This new industry would naturally require new strategies. Because humans had more time to think and communicate (and more *need* to think and communicate), language began to flourish. After a few generations the old shelters became permanent homes, and a surplus of grains quickly revamped the old system of barter into a more sophisticated method from which money — a truly symbolic system of exchange — eventually emerged.

As decades gave way to centuries, wild goats, cattle, and sheep were domesticated, and private property, a concept entirely beyond the hunter-gatherer, took hold. Populations increased, and wherever the natural elements were supportive small settlements were transformed into small towns. By

8,000 B.C., for instance, Jericho — situated on a lively natural spring that generated some 1,200 gallons of water daily — had swelled to a population of over 3,000.[6]

The old tribal identities, so vital to the hunter-gatherer's way of life, were in time subsumed in a new group: a town centered identity. Tribal chieftains and their totem rituals gave way to kings who ruled via new pantheons of gods who in turn made their history and wishes manifest through the vehicle of myth. Out of level three consciousness — in which people are far more capable of organization and coordination — the new city-state emerged, something so far beyond the wildest dreams of the hunter-gatherer that it may as well have arrived from an entirely different planet. As towns became more cosmopolitan, specialization naturally evolved. From this a class elite arose, along with all the exploitation and oppression this group has historically visited upon the less powerful and less fortunate — inequity, injustice, vice, and subjugation.

Forced by agricultural considerations, time — which for the old hunter-gatherer had always been experienced as simply the movement or passing of the present moment — would now of necessity have to include something called the future.[7] For agriculture demands this new concept, this thing we all call the future (I plant today, but must wait to harvest until a distant *future* date), which had never existed before in the human mind. As Wilber astutely notes, "With the advent of farming, men and women entered an extended world of tense, time, and temporal duration, expanding their life and consciousness to include the future. This, to say the very least, is no small achievement."[8]

For hundreds of thousands of years human beings in various forms had wandered the earth, seemingly lost and adrift, trying desperately to scrape a meager life from the hardscrabble surface of an indifferent nature. But, seemingly in the blink of an eye — at least in terms of an evolutionary time scale, all that had changed. Farming, language, symbols, kings, towns, time and a host of other advances suddenly sprang from the new, level three consciousness in the course of perhaps no more than a few thousand years, changes so startling and revolutionary they require a critical reevaluation of the natural sequence of human emergence. Things never before conceived sprouted then flourished in this dynamic, vibrant world of myth. Taken in the aggregate these phenomena took on the name we have all come to know them by today: civilization.

While all this unprecedented growth, this explosion of consciousness that resulted in civilization, was in many ways wonderful, there was a dark side to this stunning emergence as well. For in the world of psychological growth, for every potential positive there is always a potential negative lurking just offstage. Growth affords potential, in other words; and wherever there is potential there is always the possibility of misuse, of bad choices, of failure, and of regression. Recall, as well, when new stages initially emerge, they naturally emerge in their most immature form, and immaturity — at any and all levels — invariably harbors problems.

Because this initial level three period of growth was entirely new, novel and therefore untried, it seems that just as many things went wrong as right. Those mistakes have given us, and continue to give us, the downside of civilization — human sacrifice, bondage, murder, war, just to name a few. "This new arrangement unleashed on mankind regular and massive miseries that primitive societies encountered only occasionally and usually on a small scale."[9]

But as we shall see, these problems really weren't a function of civilization. They were the product of a new consciousness mishandled by those who initially stumbled upon it and were naturally unprepared for its enormous potential. Like small children learning to walk, they fell and bumped their knees as often as they succeeded in standing; and while their successes catapulted the world forward, the bumps have left their marks on us all.

Eventually a new sense of individual selfhood would emerge during this period, but that would take time and it would not be easy. This fact will become absolutely critical for the ultimate understanding of war, so it should be underlined and recalled for future reference. As pointed out in the last chapter, each level of growth has a new way of understanding or apprehending itself and level three was no different. As the level two self was a mélange of body, mind, and environment, this new level three self would be far more developed, far more aware, but not yet the distinct mental self or ego that most mature adults experience today. Because the old nomadic tribe had been lost or consumed by the organized society of the new, bustling city, and each individual now took on a specific task in this expanding civilization — farmer, baker, hunter, fisherman, craftsman, etc., the identity taken on by the level three self was that of the part or function that person performed. Life's choices, opportunities, and outlook were not determined by one's abilities as they are today in our modern world but by one's role or station.

Recall that in the previous chapter we noted how Native American hunter-gatherers' names were often a fusion of the self and the environment — Sitting Bull, Small Cloud, Running Fox, and the like. But at level three the individual had in a sense suddenly become a member of a greater cast, and the self soon came to be identified with its particular role in the overall production. And this definitely represented a more clearly developed, more specific sense of self. This is why even centuries later many level three English surnames, as an example, would emerge essentially as nothing more than descriptions of function: Taylor, Baker, Miller, Schooley, Tollman, Smith, Gardner,

Fisher, Shoemaker, etc. The role was the person, and the person was the role. The two at this level were indistinguishable. That would change even further as level four finally began to emerge, and along with it a strong mental self or personal ego. Suddenly John, Baker from Salisbury, would no longer do, and a far more personalized naming system would evolve—John M. Baker II, for example. This level four system, of course, is the highly personalized method known and used in most level four societies today.

But there was a darker, far more fearful side to this process of level three emergence. For while the greater awareness of level three consciousness certainly provided an enormous boost to life and human understanding, it also provided a greater, sudden, and probably quiet frightening grasp of *death*. In a sense, humans had for the very first time emerged from the slumber of semi-consciousness to the wonder of life in a much expanded version, but any greater grasp of life also demands a greater grasp of death. The two go hand in hand. For the first time, human beings considered deeply the grim realities of their own inevitable demise with frightening clarity, and it must have come as a profound shock. The wondrous, new, level three self realized in a sudden convulsion of horror that its newfound majesty was also entirely terminal.

There is considerable and convincing research on this point. The detailed work of Princeton University psychologist Julian Jaynes, for instance, clearly demonstrates that during this period, "While there had been earlier graves of a sort, occasionally somewhat elaborate, this is the first age in which we find ceremonial graves as a common practice."[10] The widespread use of ceremonial graves can only mean one thing: a greater comprehension of death. As Joseph Campbell points out, the hunter-gatherer had in general a far more unclear and less troubled understanding of death. "The rites and mythologies of such tribesmen," Campbell writes, "are based generally on the idea that there is actually no such thing as death."[11] But that simple understanding would soon change, and as it did something never seen before began to blossom on the outskirts of small towns around 10,000 B.C.—graveyards. One such instance is recounted in *The Human Dawn*: "Outside the temple grounds there was a large cemetery, where the dead lay with their burial articles—pottery vessels, clothing adorned with bands of colored beads, even clay figurines of the gods themselves, personified as male and female entities wearing bitumen headdresses on their lizard-shaped skulls. Each grave was a rectangular shaft, lined with mud bricks."[12]

This reaction would soon mushroom into all sorts of extraordinary temples and pyramids, dedicated to a host of gods, from Egypt to the Americas, all elaborate new strategies to try to appease this looming disaster called death. And wherever elaborate new strategies suddenly emerged to deal with death, it is a sure bet those same strategies were nothing more than humble, groping attempts by humans to deal with their elaborate and evolving grasp of their own finitude.

The prospect of death causes in humans confusion, fear, and dread. Where those concerns are handled with a modicum of maturity and philosophical transcendence, the fear of death is often muted and life is lived well. Unfortunately, a sense of mature transcendence has not often been the case in the course of human affairs, and where the fear of death is handled poorly the results can be — as our history sadly chronicles — catastrophic.

It is by no means an odd coincidence that humans have evolved to become simultaneously the most compassionate and the most murderous species on planet Earth. As shall be demonstrated, these two disparate motivations arise from the ability and the inability to deal with an expanding self, a self that by definition must ultimately die. That self began to expand most notably around 12,000 years ago in the lands of the Fertile Crescent, and it is not an odd coincidence that war and a raft of other calamities were also ushered in during this same period.

Death, the subconscious fear of death, and the inability to deal with that terror maturely are the causes of most of the self-inflicted miseries that have ravaged civilization for the past 12,000 years. *That* is exactly why war, slavery, and human sacrifice emerged concurrently with farming, poetry, language, and religion around 10,000 B.C. The disastrous repercussions of the botched side of this emergence have haunted the world ever since. Understanding just how this all went wrong will provide a greater comprehension of the most disastrous repercussion of all: war.

Four

Civilization, Eden and the Romantic Blunder

The greatest transformation in the history of the human race occurred some 12,000 years ago with the shift from level two to level three consciousness and the host of developments inaugurated by that change. The fallout from that extraordinary event, like the rumble of subterranean shock waves, still rattles the world today. For our inquiry into war it is important to explore, even if briefly, just how this event has shaped and confused much of our thinking right up to the present day.

The Old Testament story of Adam and Eve, for instance, is probably one of the most familiar mythological tales ever told, and it seems to be one of the earliest attempts to come to grips with this momentous event. Passed down verbally from generation to generation before eventually being written down by Old Testament scribes, as Joseph Campbell points out, for many years this was a tale taken as literal fact.[1] Only over the past hundred years or so have most people come to realize that the story can be understood only metaphorically, but in that sense it seems to strike an even more profound chord. The tale of Adam and Eve resonates remarkably with the actual changes that were taking place around 10,000 B.C. in the area of the Fertile Crescent, and to understand this story is to understand the enormous confusion regarding the human condition that was ushered in by this shift in consciousness. As pointed out above, this confusion remains with us still. To grasp the story properly will provide a great deal of insight into those conditions and thought processes that eventually led to the outbreak of war. For this inquiry, of necessity we must try to piece things together from events that occurred a long time ago and, as we are about to see, the story of Adam and Eve provides a critical piece of that puzzle.

Fundamentally, the tale of Adam and Eve seems a sad, melancholy story of the passing of our most distant ancestors — the hunter-gatherer — and of the rise of the new level three consciousness that would rapidly supplant the old, less conscious days when death was not yet fully understood and knowledge was not yet a word. Almost everyone is familiar with the story.

Adam was said to have been the original man, Eve the first woman, formed from a piece of Adam's rib. They lived in a time of innocence, a time before knowing. Then the serpent whispered to Eve of the special fruit of a special tree that God had forbidden them to eat, the tree of the knowledge of good and evil. The temptation was too much. Both Adam and Eve tasted the fruit, and were then banished by God from Eden forever lest they taste of a second tree, the tree of eternal life, and become godlike themselves. As the American astronomer Carl Sagan rightly notes, from that day forward humans would know death, toil, morality, shame, the domestication of plants and animals, and even murder.[2] For centuries this has been interpreted as a "fall," but in reality Adam and Eve had simply become self-aware. The price of that greater understanding was their preconscious innocence.

The story seems far more than just an old fable. Notice, for instance, how closely the consequences of the "fall" of Adam and Eve correspond to the actual features that changed around 10,000 B.C. and that were detailed in the preceding chapter. Greater psychological awareness comes with a price, and that price is greater self-knowledge. If this awareness is handled well, humans prosper with greater self-knowledge. If this it is handled poorly, the miseries multiply.

Finally God appoints a cherubim with a flaming sword to stand at the gates of Eden and guard against any return by Adam and Eve. The point here seems clear. Once humans become aware — once they have greater knowledge of the world and themselves — there can be no turning back. Semiconscious life was bliss only because for some few *ignorance is bliss*. In the modern Western world few individuals are ignorant of the toils and struggles of life (except those suffering from acute psychopathologies) and while the romantic attraction of a pre-civilized existence bereft of all the modern dilemmas at times may have its allure, few would truly wish to substitute the blissful ignorance of the bluebird or deer for the greater satisfaction of knowing. I'm afraid we have all departed Eden, that world of half-conscious wandering and innocence, long ago and forever. This is because Eden was not a place. Eden was a state of mind, and a state of mind that humanity, for the most part, has long since departed.

It is difficult for us today to imagine what it must have been like when the human condition, a way of life and state of mind that had remained virtually unchanged for millennium upon millennium, evolved so radically over a relatively short period of time. In some ways it must have seemed like the end of the world, and in a sense it was — the end of the hunter-gatherer world. Suddenly the old ways would no longer do, yet the new ways seemed fraught with all sorts of perils, problems, and confusions. So the story of Adam and

Four. Civilization, Eden and the Romantic Blunder

Eve comes down to us today full of a sad sort of whimsy, a melancholy yearning for a simpler, less complex life. Yet the story also offers the clear understanding that a single bite from the fruit of greater awareness — of the knowledge of life and death — has set humanity off on a course from which we can never return.

At the time the story was conceived, of course, no one had any real understanding of what exactly had transpired, and it would not be until the latter half of the twentieth century that developmental psychology would sensibly delineate the sequential stages of human growth. For those confused so many centuries ago, this knowledge came about 12,000 years too late. Subsequently, it seemed somehow logical that humankind had indeed suffered some sort of calamity, when the opposite was in fact the case. Greater awareness is not a loss but a significant gain. Lost in the story are the enormous benefits of increased awareness. To a very great extent they have been lost to history as well, so much so that all that seems to echo forth from the tale of Adam and Eve today are the miseries.

This unfair and unfocused grasp of the story, which concentrates on the ills that have plagued mankind rather than on the benefits of greater awareness, has been almost as damaging and costly over the course of history as the ills themselves. Ken Wilber's analysis of this problem demonstrates that two interpretations of this event forked thousands of years ago, and that those two divergent paths still reverberate through much religion, philosophy, social studies, and even ecology to this day.[3] One path suggests that the road to human progress lies ahead: to move forward, to grow as both individuals and as a species. The other suggests that the path to human freedom, dignity, and progress lies in our past: that civilization is the culprit that has robbed humanity of its rightful place, and that as a matter of pure logic it makes sense that if civilization were simply removed from the equation humanity would naturally rebound to its preferred, natural condition.

The more confused path is the latter, the belief that views civilization as the unnatural and malignant cause of virtually every human misery and suggests that the course to peace and prosperity is backward, back to our original innocence, back to Eden. This is essentially the Romantic view of things, and modern Romanticism can be traced directly to the eighteenth century French philosopher Jean-Jacques Rousseau. "The notion — which is the central tenet of Romanticism — is that we start out in a type of natural unity and wholeness, but that wholeness is fractured, broken, and repressed by the world of culture, speech, and reason. Thus our task is to *recapture* the prior wholeness and goodness."[4] Since this worldview sees civilization as the cause of all human ills, the avoidance of, or the destruction of, civilization or some aspect of civilization, is always conceived as the proper corrective measure.

And exactly there lay the rub. Since Romanticism is founded upon a false interpretation of the human condition, every attempt to "correct" what went wrong, to recapture that repressed wholeness and goodness, fails, often miserably and often with terrifying results. Because many believers accept the Romantic notion with an almost religious zeal, when their initial efforts at recapturing this natural human goodness fail there seems no alternative but to try harder, and harder, then even harder still. And when all of that fails, as naturally it must, in step the goons.

The French Revolution serves as a perfect example of this process from beginning to end. Initiated with high talk and noble goals, the revolution devolved over time into a bloodbath called "The Reign of Terror," where heads were rolled in search of this natural human goodness but only rivers of blood were found. And once the terror begins there is no natural way to stop it, because that goodness everyone is digging for always seems just one or two more beheadings away. Just ask Robespierre, the architect of the terror, who was ultimately consumed by the very madness he unleashed.

If there the matter had been laid to rest all would now be well, but that has hardly been the case. Romanticism — this need to keep digging and digging in order to uncover humankind's lost goodness — is with us still in many different varieties and flavors. Marxism, for instance, was for many years one of Romanticism's most vocal descendants, and the tragic death toll from Marxism's strident march around the globe is virtually incalculable. From the Killing Fields of Cambodia to China's bloody purification at the hands of Mao's Red Guards to Stalin's endless purges, the toll exacted in terms of life and human misery at the hands of Marxism's fanatical worshipers is certainly one of the twentieth century's saddest chapters. And here and there some indefatigable few still await the revolution of the proletariat, that wonderful day when the capitalist oppressors will be overthrown, and the workers will establish their natural, joyous, and presumably just state.

But it hardly ends there. Although generally conceived with noble intentions, aspects of today's political agenda both liberal *and* conservative, many American educational schemes and ideologies, much fundamentalist religious belief, a great deal of the teaching at Western college and university cultural studies departments, and extreme environmentalism (in particular, ecoterrorism) all float on Rousseau's pond of Romantic regression, this notion that we must go back in order to go forward. Since this approach *never* works, its adherents often become loud, strident, and ultimately tyrannical because tyranny is the only way in which they can enforce their noble objectives. Bombers (for example, the Unibomber) are almost always extreme Romantics, or at least psychotics caught up in a Romantic fantasy. It is a sad, ironic truth that Romanticism, always searching for hope, equality, and justice, generally

Four. Civilization, Eden and the Romantic Blunder

winds up devouring all three. And it consumes all three because it is chasing a lie.

It is important for this inquiry to know that the analysis of war has often fallen prey to Rousseau's notion of a buried goodness in humankind, this natural decency that the agencies of civilization have somehow managed to repress, thus creating the unnatural conditions in which men find war acceptable. In *Of Arms and Men*, for instance, Robert O'Connell's excellent review and analysis of the evolution of weapons, when discussing warfare during the sixteenth century even the usually discerning O'Connell falls prey to the Romantic fallacy and laments that "it does not seem coincidental that of all motives for mass violence except perhaps race, religious differences are most effective in undermining a sense of shared humanity and allowing men to behave as if enemies are actually of another species."[5]

The notion professed here by O'Connell is clearly that religion somehow obviates our natural inclination to love and care for one another as equals. But the fact of the matter is, the capacity to grasp all humanity as an equal is actually a rather advanced psychological ability, an ability, in fact, demonstrated by the Swiss psychologist Jean Piaget not to emerge until level four, a level it may be recalled, Piaget called *formal operational*.[6]

While many individuals around the world undoubtedly achieved that level of maturity by the sixteenth century — and some few before even that — the general average for Europe during the period O'Connell is discussing was at best level three, with a considerable portion of the population still struggling at level two. The sense of "shared humanity" O'Connell laments as missing never existed. Until very recently most civilizations historically regarded outsiders as lowly barbarians, creatures they had a perfect right to slaughter or enslave at their whim, and that view would not change substantially until the twentieth century.

Even the enlightened Greek culture, a culture that would form the bedrock of Western civilization, was ruthlessly xenophobic. Peter Green, former director of Studies in Classics at Cambridge University, explains: "Euripides held that it was proper (*eikos*) for 'barbarians' to be subject to Greeks. Plato and Isocrates both thought of all non–Hellenes as natural enemies who could be enslaved or exterminated at will. Aristotle himself regarded a war against barbarians as essentially just."[7]

Robert O'Connell was unable to detect a sense of shared humanity in the European soldiers of the sixteenth century not because that sense had been crushed by the dogmas and excitations of religion, but because it had *never been there to begin with*. It would not emerge to a significant degree until the middle of the twentieth century. O'Connell himself no doubt holds such a view of our common humanity. Many reading this inquiry may share that same view, but the vast majority of soldiers during the sixteenth century did not. In that sense, nothing had been "undermined." The sense of shared humanity O'Connell was sad to find missing had simply not yet developed to any considerable extent — a minor flaw in an otherwise brilliant book. Yet this is exactly the same mistake legions of thinkers have fallen prey to over the centuries. Searching for something in our long-ago past that we will never find will not help us understand war and will do nothing, in fact, but continually confuse and subvert the effort to understand it.

The shift from level two to level three consciousness that took place roughly 12,000 years ago was a huge event, a major leap along the road of human development that still greatly affects us today. Like the radioactive burst from a long-ago atomic blast, the philosophical fallout from this occurrence still clouds our thinking, and at times it even erupts into violence. The key to ending the blight of war lies in our future, not in recovering our past; and that is an established fact, not a philosophical supposition. In a series of studies during the 1960s and 1970s, for instance, the psychologist Lawrence Kohlberg clearly demonstrated that morality is developmental in nature, and that as we mature our sense of morality broadens considerably. The stages of morality Kohlberg discovered were essentially the same as the other developmental stages discussed herein.[8] As individual human beings we grow toward a greater depth of morality as we mature, and as a subspecies our sense of morality has been ever so slowly maturing along the same lines for centuries.

An individual's opposition to war arises as a *moral* imperative, but that sudden insight does not tend to occur until level four/five, a level that only recently has begun to emerge in the general population. The psychologist Clifford Anderson notes the following: "Maturationally speaking, our species has had a busy past century. During this time the populations of most of the advanced countries of the world have evolved from a general level of maturation that was arrested in childhood [level 3] to the point at which — for the first time — many are knocking on the door of psychological maturity [level 4/5]."[9] This fact explains why it took some 12,000 years for a concerted, international antiwar movement to emerge from the human mind, when war has been, as Robert Ardrey pointed out, the most successful of our cultural traditions. Only a century and a half ago the Confederate cavalryman J.E.B. Stuart fancied war to be one of God's greatest games. Few would endorse such a view today.

It may come as something of a shock to realize that today we are still struggling with the intellectual aftershock of an event that took place so long ago. The truth of the matter is, humanity is unfortunately not nearly as advanced as we conceive ourselves to be, and this yearning we have to return to a pre-civilized Eden is, in the final analysis, nothing more than a sad, albeit understandable, delusion. As Anthony Storr once put it, "Man's perennial capacity to imagine Utopia is exceeded only by his recurrent failure to achieve it." [10] The reason the failure to find Utopia is recurrent is because, generally speaking, we are searching in all the wrong places.

On the one hand the Romantic fallacy has often devolved into mayhem and butchery. On the other hand, the fork in the road that Wilber noted — the path of developmental growth — has been misused as well. In the long run, humankind can only hope to mature to greater depth, morality, and understanding. But in the hands of psychopathic thugs this notion has been twisted into the myth of a "super race," and the smokestacks of places with names like Auschwitz, Buchenwald, Dachau, and Bergen-Belsen have borne the awful fruit of those who were not invited along.

Call it Utopia or Eden, the hope of achieving a state of human perfection has been the dream — indeed, almost the compulsion — of mankind since humans first emerged from our semiconscious slumber. But it has been a dream utterly misunderstood, and a dream that in the hands of the morally deficient has time and again been turned into a nightmare. This confusion has at times greatly thwarted our efforts to understand war and its true genesis.

Twelve thousand years ago something of great significance occurred. Humankind moved from the hunter-gatherer level of consciousness (level two), to level three, a level that would inaugurate what we today call civilization. Much good eventually emerged from that event and, as we have seen, much that has been confused and twisted into utter calamity. War, just like dreams of Eden and Utopia, exploded across our world during this period; and just like our hopes of Eden, war was a profound misconception and a mistake that has haunted our world ever since. But civilization did not cause war.

Now that we know where *not* to go looking for the root cause of war, the search can continue for the true culprit. In the next chapter we will explore precisely how this all went wrong so many thousands of years ago, and this understanding will allow us to begin fitting the various pieces of the puzzle together to form a far more comprehensive understanding of war and, more important, what genuine prospects there might actually be for a real and lasting peace.

Five

The Psychology of War

War, along with many other inhumane practices such as human sacrifice and slavery, surfaced some 12,000 years ago during a period of *increasing* human capacity, not decreasing, as might be intuitively expected. So now it is incumbent upon this inquiry to explore the terrain of violent behavior to try to clarify just how this all went haywire. Fortunately, science has learned a great deal about human violence — both individual and collective, and if we take a close look at how humans erupt into violence, it will explain a great deal about how war emerged in the Fertile Crescent so many years ago.

Today there is widespread agreement that extreme violence is generally the result of an intolerable level of highly charged emotion that is suddenly, and uncontrollably, released.[1] That emotion, generally fear or self-hatred (which are often indistinguishable), is either directed at the self in the form of suicide, acted out in foolish ways, or projected outward onto someone else who is in turn viciously attacked or even murdered as a sort of surrogate for the self.[2]

In that sense projection is the psychological vehicle responsible for much of the violence we find around us today, and projection is a curious and fascinating thing. Anthony Storr explains: "A second feature in which we [humans] appear to be different from other animals is our capacity for projection. By projection is meant the tendency to attribute to other persons emotions, ideas, or attitudes which they do not in fact possess, but which take origin within ourselves."[3] Projection is actually so common that we all project from time to time, but it is also so slippery a mechanism that we rarely realize just what we are doing. In that projection is intimately involved with the war impulse, it is important to understand how it works. So here are a few examples.

Suppose, for instance, I have talked my way into a new job for which I am under qualified, and I know it. My lack of experience and know-how bothers me, but I need the extra pay and the new title appeals to my ego. Around the work site my coworkers talk, smile, giggle, and glance this way and that. Well, soon it dawns on me that they are all onto me, and in fact that they are actually making *fun* of me and my inexperience. As I notice this pattern more and more often, anger begins to boil within me, until one day I suddenly blurt out at two of them standing by the water cooler: "You think you're so *damn* good! Do it yourself then!" And I go storming off down the hall in a huff as the two stare blankly behind me.

Of course, they had not been talking about me at all. They hadn't even noticed my inexperience. They had simply been talking and joking about the normal things workers usually kid and joke about, perhaps a new movie or sports or someone's ugly tie. I wasn't even on their radar screens. But I blew up anyway. Why? What I felt was actually my own doubt, guilt, and self-hatred clouding then twisting my own interior perceptions until I became convinced *they*, not I, were the source of the overwhelming emotions I was experiencing. Then, instead of owning up to my genuine feelings, I had to vent my anger and frustration on *them*. The subconscious guilt and contempt I felt for myself I hurled at them, projected it outside of myself—*they*, not I, are the ones who are guilty and contemptible!

That's how projection works, and we all do it occasionally. It's very common. And the reason we project is because it is much easier to blame others for our problems and difficulties than it is to face up to them ourselves. It's as simple as that. Instead of admitting our own fear or guilt or indecision or laziness, etc., we suddenly see it in the people around us and condemn *them* for it. Here's another example.

Let's suppose I am a very petty and manipulative person, and I have manipulated my way into becoming president of the local little league, a position I have coveted for years. I accomplished this by playing up to all the people I considered worthy or important in the league for a few years, while also taking on a number of menial jobs no one else really wanted to do. Now I am president, I'm quite proud of my accomplishment, but I am suddenly aware of a new father who is helping out doing some umpiring—exactly how I got my start! My eye is on him. He has designs on my position; of this I am sure! At a picnic for the players this new father asks a simple question about league rules, and I snap at him in front of everyone, saying something like "It would do you good to be around here for *a while* before you decide to tell us how to run things!" This statement shocks everyone at the picnic, myself included. I can't explain why I blurted out such a foolish thing, and everyone is embarrassed. Once again, I simply projected my sense of self-doubt and deep-seated desires and animosities onto someone else and found him instantly guilty for all my own personal deficiencies. That is how projection works, and in this case I was both shocked and embarrassed by what came out of my mouth, the reason being I had always refused to see myself as manipulative and petty.

These simple examples provide basic illustrations of projection at work. But often the emotional charge involved with projection is far greater than these examples and the result much worse than a foolish, vocal outburst. Often projection, like a hot electric wire, conducts violence. The most common and most reported form of violence is known as the domestic dispute. This generally takes place between sexual partners. As Anthony Storr points out, where robbery is not considered a motive, the sexual partner of a victim is initially considered the prime suspect in almost every case involving violence. The reason for this is that we all create a great deal of our sense of self through our relationship with our partner, and once that important relationship is threatened, we feel the negative potential as a direct threat to our own self structure.[4]

In situations where one of the partners has had a difficult upbringing, and therefore lacks a critical amount of self-confidence, the potential loss of a partner might be felt literally as a direct threat to the very existence of the self structure — in short, a death threat. This can create an enormous amount of emotion, and the death threat might then be acted out in the form of suicide or projected onto the partner. An act of violence or even murder takes place as a consequence. It is not a random coincidence that violence often sabotages the relationships of individuals who lack self-esteem, and that murder/suicide is almost always a crime between aggrieved sexual partners.[5] The less mature we are the more we project our failings onto others, and often those failings are so fraught with highly charged emotion that they spawn uncontrollable violence.

Any way you look at it, violence can be clearly understood as a psychological mistake in the sense that the emotional charge, which is simply nature's way of coaxing a constructive change in the "self" structure of the individual, is entirely misinterpreted then violently misdirected. Instead of facing up to my own shortcomings as a mate, for instance, I might project my intense hatred or anger for myself onto my wife and attack her either verbally or physically for my own shortcomings. "No one around here will do a *thing* to help me!" I might scream at my stunned wife, because I have let some problem at work drag on too long and deep inside *I* am the one feeling ineffective. But instead of dealing with the actual problem, I get upset and take it out on someone else — in this case my wife. That sort of projection is very common. Later, realizing what a fool I had been, I apologize, and life returns to normal. But in *extreme* cases where the emotional charge is enormous, the individual might actually commit suicide or even kill someone else, and the growth process takes a dreadful turn for the worse. In such a case, of course, no simple apology will do.[6]

War arose some 12,000 years ago as a result of just this sort of misinterpretation, this sort of violent projection, when the newly emerging level three self structure was very weak and immature. So this is why an examination of the self and the violence that can be generated in the course of psychological development is exactly the place to start exploring how everything went haywire on the plains of the Fertile Crescent so many years ago.

As pointed out in chapter two, today there is general agreement that human beings develop psychologically through a number of distinct phases or stages. Since this is a natural and universal process, every human being shares the potential for enormous growth, and every human being likewise shares the potential for enormous failure. Today those people who manage their way constructively through this development with only a modicum of difficulty are said to be "well adjusted," while those who stumble somewhere along the way are said to be "maladjusted" or even "mentally ill." It is, for instance, considered normal to project your anger onto others from time to time. But if you do it a great deal or all the time, you would probably be diagnosed as paranoid or even psychotic.

Since psychological growth seems to be one of nature's goals for us, we are all under a certain amount of natural pressure to succeed. Depending upon the individual, this pressure can either be mild or extreme; but we are always under this pressure, and often it is expressed as an aggressive tendency. By this I do not mean a violent tendency, merely an aggressive inclination, a sort of need to accomplish something of value. The psychoanalyst Clara Thompson puts it this way: "Aggression is not necessarily destructive at all. It springs from an innate tendency to grow and master life which seems to be characteristic of all living matter. Only when this life force is obstructed in its development do ingredients of anger, rage, or hate become connected with it."[7]

Robert Ardrey expresses almost exactly the same insight: "Within the limits and the directions of our own individual genetic endowment we seek such a state of satisfaction as will inform us as to why we were born. We have no true choice. The force that presses on us is as large as all vital processes, and were it not so, then life would return to the swamp."[8]

For those who are well adjusted, the press of this force may be experienced from time to time as a sort of melancholy, the blues, or even a mild depression. It is the sense of feeling miserable when there is no objective reason to feel miserable, and it can be best understood as nature insisting that you still have a great deal to accomplish, when you, in fact, had no idea you had anything left to do. So you feel miserable for a while, but then you finally pick up a book, or go for a run, or go talk to your spouse, and the gradual process of growth begins anew, and the blues disappear.[9]

As previously pointed out, however, for those who are not so well adjusted this pressure can be excruciating, a force that seems so intense the individual senses the need to repress it, rather than face it directly. Ultimately, however, that pressure is simply too much to keep bottled up, and the individual faces up to the repressed material — which almost always leads to depression, but often after that healing, has what we call a mental breakdown, or explodes into some form of violence.[10] To better understand how violence can cascade into war, this inquiry will now take a hard look at one of the most extreme forms of this condition.

In today's world, one of the most feared and despised examples of this radically misshapen psychodynamic is the serial killer. Generally a male,[11] although women are also known to kill compulsively, the serial killer is usually a psychopath[12] who as a child suffered extreme abuse at the hands of someone, often the mother.[13] For years the psychopath (or sociopath) was considered to have no conscience at all, but as Dr. Theodore Rubin points out, the psychopath does indeed seem to feel and know right from wrong. It is just that his feelings and moral compass are so deeply repressed that he acts as though he has none at all.[14] Typically it is this lack of conscience or remorse that makes the serial killer appear virtually inhuman.

Often the killer will follow a pattern, or make small, intentional mistakes, or even leave behind stylized clues.[15] This is generally interpreted as an attempt to show up the authorities, or to show off, but it is actually a compulsive but minimal effort to get caught — to actually confront the real problem. But typically the efforts are generally too meager to actually work because the killer is fundamentally incapable of dealing with his own interior truth.

Ultimately the serial killer projects his self-hatred onto another innocent party and strikes in order to relieve the intense pressure he feels, usually against a lone female who seems to serve as a surrogate for the mother. Dr. Theodore Rubin explains: "Encountering precipitous hate for self is intolerable, and projection to the seeming generator of this hate, or to a person who combines both aspects can result in passionate murder."[16] The act of killing is said to provide a momentary, almost joyous burst of relief for the serial killer, and for a short period after the act the pressure for the killer does indeed seem less intense. The reason for this is that in a sense the murderer has momentarily removed or "killed" his own self-contempt by projecting it onto another human being and then destroying that person. The act of murder can thus be viewed as a subconscious but distorted attempt to rid the psyche of its toxic material. Dr. Rubin continues: "It [murder, that is] kills one's own paralyzing conscience and self-hate and displaces and projects self-hate onto victims who become representations of one's own conscience and of authoritarian figures in one's early life. Thus, shooting down an innocent victim may represent killing society, all authoritarian rules generated by society and convention, one's father and one's self-hate."[17]

But the pressure returns anew, this time with even greater intensity, for now there is even more to repress (an act of murder), and so the serial killer must strike again in a desperate attempt to relieve this additional strain. But the cycle of killing only continues to heighten his sense of anxiety all that much more. So the killer is forced to strike again, then again, and so on until either the mental walls of his world collapse, he confronts the truth of his own condition (a rarity), or he is finally apprehended. Often, the terrible truth is buried so deeply within the serial killer that it never sees the light of day, and the killer never admits to the problem or confronts his issues, which are like a safe with the truth locked away inside.

The importance of all this for our study of war is to notice closely how intense anxiety, rigidly repressed, can explode into repeated spasms of projected violence. It is also important to notice how this pressure is misconstrued by the serial killer, then projected outward onto someone else — the surrogate mother figure, in this case — who is in turn attacked and murdered in a symbolic attempt to relieve the unremitting pressure that is driving the killer to desperation. But, not only does the next act of murder not relieve the pressure the killer feels, it actually increases it. In a very real sense he is running from the truth, trying to literally "kill it," but the truth just keeps growing in intensity.

This almost suffocating sense of pressure does not, of course, force the psychopath to kill. That is simply his terrible choice. The pressure is in fact only the killer's internal psychological dynamic trying to force him to face the truth of his own dire situation, deal with the repressed material he is desperately trying to avoid, and return to the normal path of developmental growth.

But the serial killer does not have the *courage* to face his own depravity (or perhaps what he conceives as his own depravity, quite possibly childhood sexual abuse, as an example), and thus his choice to kill can clearly be seen as a profound mistake, or a weak choice, or indeed a cowardly act. And if you will recall that in chapter one of this book we cited Thomas Mann's famous remark that "war is only a cowardly escape from the problems of peace,"[18] then perhaps you will begin to get some sense of how closely war and psychopathology are related.

With all that in mind we will now return to our discussion of war and to the conditions that prevailed during the 10th millennium B.C. around the time war first exploded across the face of the Fertile Crescent. Recall that the most significant occurrence of the era was the shift from level two to level three consciousness, a shift that ushered in a greatly expanded sense of awareness for the subspecies *Homo sapiens sapiens*. This awareness allowed for an enormously greater grasp of both the exterior and the interior worlds the individual human beheld. In short order the new exterior awareness would conceive of such things as agriculture, metallurgy, architecture, the wheel, astronomy, great cities, irrigation systems, and much, much more.

From the newfound interior awareness would spring symbols, literature, mathematics, poetry and the first forms of formal religion, along with a host of other wonders. Out of the old, hunter-gatherer self—a very meager self that we have seen was confused and fused with the environment—would emerge a far more structured, stable self, but a self that was not yet all that it is today. While that interior sense of self was new and in many ways grand, in other ways that newness meant it was also young, immature, and subsequently fragile.

And as that level three self beheld the extraordinary wonder of its own stunning interior for the first time, and grasped with a sense of awe all of the remarkable mental insights just mentioned above, it would also have had to have grasped in a much deeper sense than ever before a profoundly threatening element—death. "To become aware is to become conscious of the perpetual threat of nature to the security of the self and of the inescapable fact of death."[19] The new interior awareness stumbled directly upon a far clearer comprehension than ever before of its own mortality, and that sudden comprehension had to be nothing short of terrifying. Never before had human beings grasped quite so clearly the inevitable fact that they were going to *die*. This, for humankind, represented an enormous new fear, and as Rush Dozier, Jr., points out, "Fear is a powerful organizing principle for both the individual and society. Neither individuals nor societies can ever be truly understood unless we understand their fears."[20]

There are essentially two ways of dealing with something that terrifies us. We can face it directly, or we can pretend we're really not scared—go whistling past the graveyard, if you will. If I pretend I am not scared then I repress my fear entirely (and become anxious and depressed), act out in odd ways (pick fights, for instance, for seemingly no reason at all), or project my fear onto others (*they* are the one's who are dangerous and need to be locked up). Evidence from most cultures during the period of transition from level two to three suggests that the new terror of death was dealt with in all these ways.

The archeological record from this era, for instance, appears literally consumed with images of death, so much so that it is impossible to dismiss as anything other than enormously revealing. Here is acting out on a virtually Herculean scale, and here is a big piece of the puzzle. The towering pyramids that suddenly blossomed, from Egypt to the Americas and all constructed at an extraordinary cost in lives, labor, and finance, bear clear testimony to the compulsive need humans suddenly experienced to deal with this newfound terror. "And there, in a phrase, was the grandeur that was Egypt: the mortuary cults, the Pyramid Age, the mummies, the golden death masks."[21] Today we are amazed by the scope and sophistication of the *engineering* of these monuments, entirely unmindful, it seems, of the sense of psychological desperation that brought these gargantuan structures into being in the first place. How else, for instance, can we explain the fact that all humanity seemed to share this remarkable compulsion during this particular period in history? Yet it is a compulsion that has long since passed us by. Yes, we still build pyramids in a sense, but today our Herculean engineering efforts are dedicated for the most part to commerce, not death.

The great tombs and pyramids were one way level three consciousness dealt with the fear of impending death, but any fear capable of raising the great pyramids also had to be capable of affecting the new, fragile self in other ways as well. Not only was the death fear acted upon, the terror that was death also soon became largely repressed because it was simply too fearful to deal with. Yet any fear that tremendous could not remain subconsciously dormant for long. Unable to repress such a frightening truth, the level three consciousness soon — just like our serial killer — projected death outside of itself. Ken Wilber explains: "It is at this stage [level 3], by the transformation of passivity into activity, that the fateful extroversion of the death instinct outward onto the world in the form of aggression takes place."[22] War, murder, and human sacrifice exploded suddenly onto the scene. Wilber captures it precisely: "And it is this death impact that is extroverted, at the membership level [level 3], into the peculiarly morbid, vicious, and unmitigated form of aggression known only to humankind."[23]

Just as with the serial killer, the internalized fear of death was projected outward by the fragile, level three self, and the sudden urge to kill became all consuming. "The original death terror becomes death-dealing, and *there* is the human source of joyous murder."[24] And because of the relative immaturity of the level three self and the sheer ferocity of the emotion involved, when it exploded it exploded with almost unimaginable violence. Suddenly the Fertile Crescent, a land that had enjoyed essential peace for hundreds of thousands of years, was literally awash in human blood. Recall the boast of the Assyrian leader, Ashurnasir-pal from chapter two:

> I cut off their heads. I burned them with fire, a pile of living men and of heads over against the city gate I set up. Men I impaled on stakes. The city I destroyed, devastated. I turned it into mounds and ruin heaps, the young men and maidens in the fire I burned."[25]

Robert O'Connell captures the brutal essence of these early campaigns, suggesting that the victor of these clashes often turned them into orgies of virtually incomprehensible slaughter.[26] The frantic terror of death projected outward becomes in turn the compulsive need to slaughter. And it swept across the Fertile Crescent like a tidal wave. Stanton Coblentz had this to say: "It is a thought, a desire, which, founded upon a magical belief, has been converted into a devouring passion, and thunders across the cities and civilizations with the blindness of an avalanche."[27]

The level three consciousness stumbled headfirst into war simply as a means of staving off death in precisely the same sense that our serial killer murders again and again to stave off his fear of the truth hidden secretly within himself. And just as with the serial killer, the slaughter comes with a sense of almost festive liberation.[28] In *War Is a Force That Gives Us Meaning*, Chris Hedges gets it exactly right when he states with brutal simplicity that "war finds its meaning in death."[29]

The massacre of others momentarily appeases the frightful specter of death that was initially grasped then repressed, and so the projected death terror turns into a literal frenzy of bloodletting. The more slaughter is engaged in, the more momentary release is felt, and from this momentary release a sense of immortality arises.[30] The more the enemy suffers, the more intense is our sense of immortality.[31] And that suffering would cause rivers of blood to run through the golden grains of the Fertile Crescent and on across the globe for thousands of years thereafter.

The enemy is simply any distant town, tribe, or city — anyone *else* will do. The more the enemy is killed, and the more savagely it is done, the more momentary relief is achieved, and that great sense of relief breeds an odd sense of immortality. This is the predatory compulsion this inquiry has been searching for, natural aggression driven to a frenzied state of excitement by the terror of death, which is then projected outward. Wilber believes "whatever natural aggression is innately present in humans, the important point is that it is amplified through conceptual domains, and that amplification — itself *not* genetic — constitutes the special, morbid, excessive aggression known only to man."[34] Wolves do not war because wolves do not grasp the trajectory of their own lives and thus the inevitably of their own deaths. Suddenly, humans did, and our bloody history from that point forward speaks for itself. The Ngogo chimpanzees (discussed in chapter one) might fight upon occasion but they do not war, and they do not war because they are not developed enough to grasp life's implications. War is a *human* construct, and a construct that arrives at a very particular moment in the development of *Homo sapiens sapiens*.

It is not an odd coincidence that war and human sacrifice emerged more or less coincidentally. What, after all, is human sacrifice if not a fumbling, pathetic attempt to stave off death by means of the ritual, substitute butchering of *others*? Someone must die, I know (is our reasoning), but if I slaughter hundreds or thousands of humans in ritual form, perhaps I will be spared to live and kill another day. That reasoning is the essence of all human sacrifice. Death no longer controls me; now I control death. Genocide, humanity's gift to the twentieth century, is simply human sacrifice shifted from the fumbling stone implements of the Aztecs to assembly-line, death-dealing automation — like the production of automobiles or hair gels — a more modern, mechanical, and therefore efficient approach. Human sacrifice, just like human warfare, is clearly an attempt to deal with the enormous psychological impact of death by frantically killing someone *else*.

But as we have seen, the consciousness at this emerging level grasps not only death and war, but additionally a vast new range of constructive human potentials. And this is exactly the reason why humans became almost overnight the most creative, sophisticated, compassionate, *and murderous* creatures on earth. O'Connell notes this curious dichotomy: "The idea of warfare as the prime motivating force and major preoccupation of emerging urban societies must strike some as entirely overdrawn. Generally, the appearance of civilization is perceived as having broadened and enriched life, not simply militarized it."[33]

In the first chapter of this inquiry it was noted that any explanation of predatory warfare had to answer why war and civilization arose simultaneously; why the more rather than the less developed creature initiated war, and finally what force caused the drive to war to evolve over the centuries. The first two of these questions have now been answered. The third will be addressed in upcoming chapters.

The riches of civilization would continue to grow, while the ravages of war would leave their enduring and indelible scars on the architecture, memories, and psyche of humankind for thousands of years to come. The level three world would become a world virtually dedicated to warfare. "The image which emerges from monuments, inscriptions, and the writings of contemporaries is that of a dangerous world where attack could come from any quarter and only the militarily strong survived."[34] This was a period in human affairs of almost constant warfare. Virtually every history from this human epoch resonates with war, conquest, and slaughter.[35]

The subconscious urge to immortality through slaughter: this, then, is level three war, and to a great extent, the level three world. War seems to come out of nowhere and for no particular reason that the *logical* mind, at least, can decipher, because war, as we have now seen, is not logical in that sense. It is logical only as an extreme psychological reaction to an extreme psychological fear, that of death. A far more conscious grasp of death comes to the level three mind just at this moment during the long ascent of humankind; and just here, as well, come pyramids, human sacrifice, bondage, and war. Those facts are by no means an odd coincidence. Indeed, they explain a great deal of our common history that previously has been considered essentially inexplicable. Slip those puzzle pieces together, and the sudden, violent emergence of war during this particular historical period begins to make sense — just here, just now.

So it would be for thousands of years to come, city-states and empires, both large and small, swallowing one another whole, until finally a new wrinkle will be added to the mix. This new wrinkle will bring with it new technologies for even greater terror and destruction, just as it will inaugurate new wonders; and it would change the face of predatory warfare for thousands of years to come. What could cause such a thing?

The answer is the first, tremulous emergence of level four consciousness.

Six

The Archetype, the Ego and the Genesis of War

As has been demonstrated, with just a little insight it is easy to understand that our ancestors left behind them an abundance of clues confirming the ragged course of their psychological growth. These clues remain embedded in their art, weapons, architecture, lifestyles, myths, and so forth. With just a little understanding of the human developmental curve, they become easy to pick out. We have seen, for instance, how the totem pole represented a fusion of self and environment for the level two hunter-gatherer, and it has also been demonstrated that city walls, pyramids, weaponry, and chronicles of slaughter all clearly affirm the emergence of level three consciousness sometime around 10,000 B.C.

None of this is terribly surprising. Each shift up the line of development represents almost a quantum leap in conscious capabilities, and each stage rather naturally tends to grasp and represent the world as seen and experienced in that level. In just that sense the emerging level three mind was able to grasp and produce "civilization" in a remarkably short period of time, something the old hunter-gatherer had not come even close to conceiving during the course of thousands of years of wandering. This was no small, evolutionary development, by the way, but a virtual geometric explosion in abilities, and the results remain for all to see.

The hunter-gatherer had thought of himself as part and parcel of the environment around him, and his cave paintings, rituals, and beliefs mirror perfectly that embryonic fusion. The level three consciousness, on the other hand, produced language, architecture (on a grand scale), written art forms, religion and philosophy, and it grasped its "self" essentially along the lines of the individual's role in society. For the first time human beings devised a division of labor. Thus priests were priests, farmers were farmers, and the new class of warriors conceived themselves simply as warriors. But in that sense, as well, war became enshrined during the level three experience, and the warrior became an ikon, or an archetype — to use Carl Jung's terminology — that is still very much with us today. War tumbled out of the dark side of man's psychological development in a sudden, violent burst; but it would not take long for warfare to become considered a natural condition of the human experience, and culturally enshrined as such. This process was not unique to warfare, but it certainly had devastating consequences for humankind.

The vast majority of the everyday human experience is based upon repetition, on the patterns of life laid down before us and that are repeated endlessly, at times with minor alterations, for better or worse. I may, for instance, decide upon which pair of shoes I will wear today, but rarely will I decide whether I am going to wear shoes, or clothes, or a hat when it snows. Those things I do without contemplation. The way we speak and address one another, the courtesies and rhythms of human life, the roles we fill have all been predetermined. Only rarely do we awake to take notice of the fact that we are essentially sleepwalking through life, simply following a trail that was blazed long ago. That is why people say that the more life changes the more it stays the same, because only the surface really changes and that but superficially. This is not necessarily a bad thing, because many of the rhythms are tried and true and they get us where we are gong without much thought, much like the bus we are used to taking. Rhythms and patterns allow us to live without having to reinvent the world every day, but over time they can also rob us of thought, spontaneity, even reason. We humans think of ourselves as rational beings, but very little that we do is based on true rational thought. Most of what we do we do because that is the way we have learned to do it; what we learned is simply a pattern created long ago and repeated, with a few updates, across many, many generations.

Our lives are therefore made up of basic patterns, models, they might be called, of what it means to be human. The earliest of these, those patterns laid down at the very dawn of the *Homo sapiens sapiens* epoch (or that which preceded that epoch, according to certain theorists), are what the psychologist Carl Jung called archetypes. According to Jung, many of the earliest and most fundamental of these archetypes are deeply embedded within all our unconscious minds. These are forms that give shape and meaning to our lives that arise from what Jung called a "collective unconscious."[1] Jung explains:

> And it was no less authority than Hermann Usener who first recognized unconscious preformation under the guise of "unconscious thinking." If I have any share in these discoveries, it consists in my having shown that archetypes are not disseminated only by tradition, language, and migration, but they can rearise spontaneously, at any time, at any place, and without any outside influence.... The far-reaching implications of this statement must not be overlooked. For it means that there are present in every psyche forms which are

Six. The Archetype, the Ego and the Genesis of War

unconscious but nonetheless active — living dispositions, ideas in the Platonic sense, that preform and continually influence our thoughts and feelings and actions.²

For Jung the collective unconscious represented a sort of repository of all human experience that could be tapped into and shared by all human beings. While many other theorists have agreed with Jung on the existence of the archetypes, there has been considerable debate on their nature and genesis. Ken Wilber explains:

> This phylogenetic or "archaic heritage" includes, according to Freud, "abbreviated repetitions of the evolution undergone by the whole human race through long-drawn-out periods and from prehistoric ages." Although ... Freud and Jung differed profoundly over the actual nature of this archaic heritage, Freud nevertheless made it very clear that "I fully agree with Jung in recognizing the existence of this phylogenetic heritage."³

Piaget, on the other hand, differed with Freud and Jung in that "he does not see the archetypes themselves as being directly inherited from past ages, but rather as being the secondary by-products of cognitive structures which themselves are similar wherever they appear and which, in interpreting a common physical world, generate common motifs."⁴ Thus for Piaget, it was the emergence of the varying stages of psychological growth themselves that caused the varying experiences and interpretations called archetypes, and not any preexisting, archaic, universal heritage.

Regardless of the genesis of the archetypes, one thing remains clear for our inquiry into war, and that is that early human experiences laid down certain patterns or models and that these have had a profound influence on human activity ever since. Again, Ken Wilber explains: "In other words, the Jungian archetypes are for the most part the magico-mythic and 'archaic images'— they should really be called 'prototypes'— collectively inherited by you and by me from past stages of development, archaic holons now forming part of our own compound individuality."⁵

The point of all this for our inquiry into war is that as level three consciousness began to arise, and as civilization therefore began to develop, many new archetypes would have imprinted themselves onto the human experience, and there is little question that one of those new archetypes would have been the warrior. Thus was war over the centuries slowly enshrined as a natural feature in the affairs of *Homo sapiens sapiens*. As Dr. Robin Robertson, an expert on Jungian psychology, explains, "There is no way to decide how many archetypes there are. There are seemingly archetypes for every person, place, object, or situation which has had emotional power for a large number of people over a large period of time."⁶ While at first this may appear to be a staggering number, all Robertson really means is that the groundwork for human development and understanding of the world originated with basic patterns of thought, and those patterns have been repeated, molded, and improved upon over the centuries. In that sense there can be little doubt, when considering the prevalence and sheer emotional impact that war has had on the human experience, that the warrior became an archetype early on, and war a powerful aspect of human culture.

A formative consistency led to habit, and habit over the millennia eventually formed the ruts of war, an ingrained cultural response. Precious little thought was involved. Again, we humans tend to see ourselves as rational beings, but much of what we do has little to do with logic and everything to do with habit. Our experience with warfare was certainly no different. Robert Ardrey says, "No man can regard the way of war as good. It has simply been our way. No man can evaluate the eternal contest of weapons as anything but the sheerest waste and the sheerest folly. It has been simply our only means of final arbitration. Any man can suggest reasonable alternatives to the judgment of arms. But we are not creatures of reason except in our own eyes."[7] As explained in chapter five, the original genesis of war was clearly psychological, the immature projection of the death threat onto others; but once the warrior had been established as a basic model of the human adventure (an archetype), and war legitimized as a result, warfare became a compelling, knee-jerk response to situations invested with high emotion. War became a ghastly habit.

This conclusion may at first seem superficial, but after studying and writing about war for decades, historian John Keegan arrived at exactly the same conclusion, and his thoughts regarding this are worthy of consideration:

> All that we need to accept is that, over the course of 4000 years of experiment and repetition, warmaking has become a habit. In the primitive world, this habit was circumscribed by ritual and ceremony. In the post-primitive world, human ingenuity ripped ritual and ceremony, and the restraints they imposed on warmaking, away from warmaking practice, empowering men of violence to press its limits of tolerability to, and eventually beyond, the extreme. "War," said Clausewitz the philosopher, "is an act of violence pushed to its utmost bounds." Clausewitz the practical warrior did not guess at the horrors toward which his philosophical logic led, but we have glimpsed them. The habits of the primitive — devotees themselves of restraint, diplomacy and negotiation — deserve relearning. Unless we unlearn the habits we have taught ourselves, we shall not survive.[8]

Restraint, diplomacy, and negotiation are indeed the fundamental ingredients required for a rational approach to human disagreements, but here Keegan tumbles headfirst into the romantic fallacy himself. We will not discover those qualities by attempting to recover our primitive past, by rummaging around in pre-civilized societies in search of the tools modern civilization will require for a peaceful tomorrow; we will find them only by forging ahead into a level four future where those qualities become the natural condition of the human outlook, and negotiation, not war, the natural impulse to the settlement of disputes. As has been pointed out, in many areas across the globe this process has already begun to take hold as the level four mind evolves toward higher morality and democracy and its natural inhibitions against predatory warfare. With this, Wilber (for instance) appears to be in complete agreement:

> The previous, mythic-imperial [level 3] empires (East and West, North and South), with their inherent dominator hierarchies, were deconstructed by the rise of egoic-rationality [level 4] (wherever it did arise) and the switch from a role identity to an ego identity, and the correlative rise of the democratic state, where dominator hierarchies were replaced, in law, by (1) free and equal subjects of civil law, (2) morally free subjects, and (3) politically free subjects as citizens of the democratic state.[9]

But, in that many governments and institutions in the world today remain firmly in the grip of level three patterns of thought, the urge to predation remains a constant problem. Again, Ken Wilber elaborates: "There still exist, of course, mythic-imperialisms [level 3] that wish to dominate the world — that is, imperialisms that will accord you the status of equal world citizen *if* you embrace the particular mythology (with its particular dominator hierarchy). If you don't want to embrace the mythology, they will be glad to help you do so."[10] In that sense war today remains a level three habit, but habits can be modified and eventually broken. As humanity struggles into the twenty-first century and a far more developed world, the opportunity clearly exists to modify the old warrior archetype that blossomed thousands of years ago and that has haunted human existence ever since.

That original warrior archetype gave us level three war, which was by and large parochial, predatory, impulsive, and fixated on blood and booty. These wars had nothing at all to do with *ideas*. Indeed, the level three warrior was in essence scarcely distinguishable from the latter-day pirate or buccaneer. These early conquests were generally ill thought out and tactically the equivalent of a modern street brawl or gang fight, only on a much larger scale. Robert O'Connell explains: "Here was the key to this kind of combat — find and kill the leadership. Thus, military historians miss the point when they provide detailed tactical descriptions of such actions, imputing motives which did not exist and labeling as maneuver what really only amounted to crazed looting. Such engagements turned on one encounter only, the melee around the headmen. Should the king and his prime retainers fall, the battle was lost."[11]

These sorts of campaigns were exemplified by the conquests of the Sumerians, Akkadians, Gutians, Hittites, and Assyrians — short, brutal, decisive. Wielding new, far more effective metal weaponry and body armor, mobilized into tactical units predicated upon the number ten, with an array of archers, spearmen, and chariots, even the first embryonic use of cavalry, the Assyrians, for instance, would establish an empire that would last some seven centuries, yet an empire that — like all those it had brutally absorbed — would crash into sudden and violent oblivion.[12] Often wars were determined by the outcome of a single, bloody confrontation. On far more than one occasion vast empires that had been hammered together by predatory efforts over hundreds of years were swept away in a single afternoon.[13]

The endless fighting and dreams (and fears) of conquest this human epoch is noted for would leave their marks on the architecture, literature, and psyche of *Homo sapiens sapiens* for centuries to come. "It is probably safe to assume that up to A.D. 1650 more resources were invested in defensive walls than in all other public works combined. Regardless of culture or politics, virtually every urban center above hamlet size in the temperate band of Europe, Asia, and North Africa would come to have its surrounding ramparts."[14] In China the Great Wall would be constructed to serve the same purpose, only on a far more colossal scale. So extraordinary was the effort that today it remains one of the most striking manmade features on earth, visible even from an orbiting spacecraft. Around 2,000 B.C. a gradual shift began to occur, until approximately 500 B.C., when that gradual shift exploded into a crescendo of change. The infant emergence of level four consciousness, and the world most of us inhabit today, had begun to take root.[15]

Level four would quickly produce a distinct culture of its own, albeit much of it founded upon the original level three prototypes. In that sense, art, architecture, philosophy, warfare, religion, and the like were not re-created but would be taken instead a step further — rethought and amplified. Thinking would slowly shift from literal to rational (what we call logic), and ultimately modern science would emerge as a product of that growth. The scientific method, the verification of a hypothesis by means of exhaustive and repeated *testing* would eventually give rise to an entirely new approach to nature, and modern science would be born. But all of this was, of course, at the time of the initial level four emergence, still far in the future. Humans for the first time became truly self-conscious, aware of their own inner thoughts, suddenly capable of deep introspection and self-analysis. As if awakening from a dream, human beings began thinking deeply for the very first time. A whole new world had suddenly dawned.

The modern self — the self most of us currently experience — would also appear, with all its new, remarkable abilities: abstract, logical, and introspective thinking along with hypothetical reasoning.[16] From this would come the ability to conceive what *might* be and the capacity to take the place of the other, from which would arise idealism and a far wider, deeper grasp of morality.[17] Notions of free will and personal responsibility would make their debut. And right along with those remarkable abilities would come the incredible arrogance, self-absorption, and numerous other pitfalls inherent in this new and immature level four ego.[18]

The first glimmers of this shift blossomed slowly (2500 B.C.–500 B.C.) because the conditions level 4 consciousness required were confined to a very limited group — the upper class of wealthy and powerful leaders. In short, it would be the kings and their courts who would lead the way into this new phase of human consciousness, and since a king and his retinue were generally warriors, for humanity in general this would prove to be a dicey proposition.[19]

In other words, the working model for level four consciousness would be the warrior king, and the results of this marriage would only serve to amplify war. Wilber explains: "The warrior kings, in short, cut themselves off from subservience to community. Instead of serving society, they arranged the reverse: a replacement of social sacrifice with undiluted personal ambition."[20] In that sense, the newly emerging ego, born to war, would in turn seek validation and growth through predatory warfare. The stage had been set for a disaster of unmitigated proportions. All that was needed were the players. Those players began to arrive on the scene around 500 B.C., and remarkably, so did their story — a new story. Suddenly, and right on cue, the great myths of the era took a remarkable turn. The hero myth was born.

Anthony Storr explains:

> If one examines any typical hero myth, one finds the same kind of story told over and over again. A child or young man, often the youngest or the most deprived of the family, sets out on a journey to make his fortune. He is often ridiculed by his family, who do not recognize his potential. The journey is attended by many dangers, and the hero may find help from supernatural or animal sources. Usually the hero has to kill a monster or perform some dangerous task, often with the object of freeing a lady in distress whom he wins as bride. It is reasonable to regard such stories as allegories of the long process of growing-up.[21]

The new hero myths were, at their most basic level at least, not about origins, or conquest, or even the gods at all. They were about the process of *growth*. They were the story of the new ego trying desperately to break free from the shackles of level three domination. The quest for fortune was in fact the quest for maturity, and the monster that was slain was the old, level three identity that had to be cast aside.

These stories were representative of an entirely new phenomena, and they were new for a reason. "The first aspect is that *the Hero is simply the new egoic structure of consciousness*," Wilber explains, "which, coming into existence at this time, is naturally given living expressions in the mythology of this period."[22] In short, the new level four consciousness had an important story to tell, the tale of its own, desperate struggle. Wilber makes it clear: "And the true hero myths do not emerge before this period because there were no egos before this period."[23]

Far from fairytale or fantasy, the hero myth, properly understood, is actually a story of heroic development, of heroic struggle. These myths are tales of the journey of the level four ego overcoming the clinging obstacles of its past — all the lower-level structures that were trying to inhibit it — and flowering into a new, dynamic form of human consciousness. Ken Wilber continues: "Now the second important aspect of the egoic Hero Myth is the nature of the monster that is slain, captured, or subjugated.... The dragon guards the ego — and that's what the Hero must liberate."[24]

By slaying the dragon the ego finally emerges, free and unfettered. In many ways the hero myth is the tale of the birth of our own world, and that is the reason these stories have held people spellbound for centuries. Gilgamesh, Jason, Indra, Lancelot, to name but a few, are *us*, their struggles to free themselves from the dragons of the past are our own struggles. Many modern stories and movies like *Rocky*, *The Natural*, and *Shane*, are nothing more than these same ancient hero myths dressed up in modern garb.

The greatest storyteller of the level three era devoted himself entirely to this new phenomena. His name was Homer, and his *The Iliad* and *The Odyssey* are widely read even to this day. In these tales the new egoic structure seems to lift its head, open its eyes, and recognize "self" for the very first time. Listen to Achilles, for instance, as he preens before a beaten adversary in *The Iliad*, just outside the walls of Troy: "Do you not see what a man I am, how huge, how splendid?"[25]

Such is the vanity of the nascent ego, and this is precisely what the psychologist Jean Piaget discovered when he studied the emergence of level four consciousness in modern adolescents.[26] Over time this extreme egocentrism (which is a hallmark of the developing ego) will give way to a far more mature individual, but the initial emergence of this stage is one of enormous self-involvement, simply because the self, for the first time, has been discovered. Today, boasts like that of Achilles are generally heard only from the mouths of immature adolescents or perhaps professional athletes, but such vanity was common in the early hero myths just as today it is common in the early, level four development of the individual. F. Philip Rice says this about the emergence of level four awareness in today's adolescents:

Another effect of adolescents' intellectual transformation is their development of a new form of egocentrism. This egocentrism is manifested in two ways: through the development of what has been termed **imaginary audience** and personal fable ideations.... They direct their thoughts toward themselves rather than toward others. They become so concerned about themselves that they may conclude that others are equally obsessed with their appearance and behavior.[27]

Remarkably, Robert O'Connell discovered signs of this egoic emergence while exploring this time period, but from an entirely different perspective. O'Connell's work was focused not on psychology but on the relationship between war, humans, and their weapons. Yet he noticed a distinct change in how combatants went about their business around the period of Homer's writings. He realized that combatants were for the first time fighting in accordance with freshly devised systems of conduct or honor, away from women and children, and under a demanding code of combat.[28]

This is also the first time when what O'Connell terms the urge to close would be displayed. Here warriors take to the field armed for close combat, the preferential form of which is to advance to close quarters with the adversary and fight man to man. This Homeric urge to close, O'Connell argues, would in subconscious form dominate Western military tactics — on land, sea, and eventually in the air — for thousands of years.[29] It is the cavalry charge, the infantry advance, the broadside, and eventually the dogfight in the skies. Above all else, Homer's writings made war *heroic,* and predatory warfare the specific vehicle through which that heroism could be realized. In that sense, the ego would evolve a code of warfare, and Homer would stamp that code, the warrior archetype, with the heroic imprint. It is an imprint that lives with us still. Lastly, it was O'Connell's judgment that these codes of honor were not simply the work of Homer's creative whimsy. They were entirely authentic of the era.[30]

But it is apparent also that codes of conduct are really nothing more than the behavioral demands of the new ego, a consciousness that will no longer simply "have at it" as in days of old, but must now fight in a prescribed manner that it considers appropriate for its lofty status. Suddenly style is every bit as important as success, and from this will soon flow the painstaking and meticulous sense of honor and fastidious decorum that will dominate military culture worldwide for centuries. History certainly demonstrates that the egos of warriors, almost like those of ballet dancers, can prove very delicate instruments, indeed. It is often noted, for instance, that during the American Civil War Robert E. Lee had almost as much difficulty keeping his top officers out of duels with one another as he did in fighting the entire Federal army. The ego, as Lee discovered, can be very prickly thing.

While Homer's *The Iliad* describes certain new mores and codes of conduct, *The Odyssey* is his real declaration of level four independence. Ulysses, the hero of the saga it will be recalled, must return to his home in Greece after years away during the Trojan War. The story — precisely as outlined by Anthony Storr — involves a number of fantastic encounters Ulysses and his crew must endure on their journey home. Many do not survive, but Ulysses does, and he returns home a new man. While Ulysses is certainly a skilled warrior, he succeeds as much by wit, craft, and pluck as he does by might. These traits — just as with size, speed, beauty, and strength — will become the celebrated traits of the new, emerging ego.

Homer's tale of Ulysses has thrilled readers for over two thousand years, and it connects even now with modern readers because it speaks directly to our own inner journey. We all must discover our own egos, find our "selves," and for most people today this generally takes place during adolescence or the years leading into early adulthood. During the course of the normal human developmental curve, we all must struggle our way out of level three to level four, subjugating as we go the demons and demands of our mythic, role-oriented past.

In a sense *The Odyssey* is a script written for the heroic journey, but in the end it remains only a story. The real journey would be made in flesh and blood. Over the centuries many attempts would be made to fill Ulysses' shoes, and many of those attempts would provide sad testimony to the needless carnage that can reign when the newly emergent ego becomes infatuated with itself and the misbegotten notion of triumphant expansion through heroic, predatory warfare.

Some of these egos would become so bloated that only the conquest of the entire known universe seemed sufficient to satisfy their boundless lust. Far more than once has the world trembled under the boots of their armies and the fists of their henchmen. Darius, Tamerlane, Hannibal, Caesar, Charlemagne, Genghis Khan, Napoleon, Hitler — these are just a few of the names of the men who have attempted to find themselves through predatory warfare. Many, many others would raise the sword.

But only one man would come to truly personify the reckless swagger of this new heroic ego cast in predatory garb, a soul lost, in a sense, to conquest ad infinitum. For thousands of years his legend has inspired legions of imitators, and like a dull drumbeat his deeds still echo in the war-scarred halls of Western civilization. Like a siren's song, his name alone seems to invite emulation. Indeed, it is said that Napoleon invaded Egypt for no other reason than to follow almost blindly in his footsteps. Stanton Coblentz makes this observation: "He has been something more than a man; he has been a divine hero, a legend; he has been worshiped not for what he was but for what men imagined him to be."³¹ His life serves as a striking example of all that can go right, and all that can go wrong when this new, level four ego confuses true development with martial conquest. Born to King Philip in the backwater state of Macedonia in 356 B.C., this man became the greatest and most celebrated warrior of his time, perhaps the greatest the world has ever seen. He is known to us today as Alexander the Great, and an examination of his life and deeds will reveal a great deal about the next significant leap in the evolution of war.

Seven

Alexander and the Warrior Archetype

It is late spring, 327 B.C. Alexander of Macedon watches with pleasure as his army, the most successful and feared military force ever assembled, slowly navigates the Hindu Kush and descends through the frigid mountain passes into India. They have crossed the dreaded highlands of present-day Turkestan and Afghanistan, and now the open Indus Valley — the doorway to the Far East — beckons. Alexander has divided his force, sending half ahead through the Khyber Pass while he personally leads the second portion on a round about trip through the Kushan Pass then up the Kunar River in order to cover the main column's left flank. If all goes as planned, both columns will soon rendezvous on the Indus River.[1]

Eight years ago Alexander's father, Philip, had been treacherously murdered while leading his Greek forces east toward Asia. Alexander, only twenty years old at the time, immediately inherited not only his father's crown but also his enormous ambitions, and in those intervening eight years Alexander has conducted the most extraordinary military campaign the world has ever seen.

His force is both remarkably capable and sophisticated, meticulously assembled to meet virtually any contingency. Alexander has at his disposal heavy cavalry, light cavalry, archers, slingers, dart and javelin throwers. His siege train consists of lightweight catapults that can be rapidly assembled by his engineers, as well as battering rams and portable siege towers.[2]

But the heart of the Greek force is the infantry. Called the Foot Companions, the infantry consists of hoplites (armed spearmen) recruited from the middle and upper classes of Greek society. Armed with a small shield and short sword for close-in fighting, the hoplites' principal weapon is a *sarissa*, or lance, some 14 feet in length. Highly trained and highly motivated, fighting in the famed Greek phalanx sixteen rows deep, the Foot Companions have been able to cut their way through every opponent they have faced thus far, often inflicting severe casualties while suffering very few themselves.[3] Trevor Dupuy says, "Careful organization and training programs welded the mass into a military machine which, under the personal command of Philip and later Alexander, probably could have been successful against any other army raised during the next eighteen centuries; in other words, until gunpowder weapons become predominant."[4]

Indeed, up to this point the Greek forces under Alexander have proven to be almost invincible since their landing in Asia Minor. Alexander, always leading from in front with the light cavalry — or Horse Companions — personally leads the Greeks into battle, and the effect of his leadership has been critical. In 334 B.C. the Greeks decisively defeated Darius III along the Granicus River, systematically slaughtering thousands of Persians while losing only 115 of their own.[5] The following year Alexander marched into present-day Syria and again faced Darius. This time the Persian emperor waited near the town of Issus with a force more than twice the size of Alexander's. Once again the Persians were routed from the field.

The victory at Issus was followed by the siege and utter destruction of the port city of Tyre — where Alexander had over 2,000 defenseless inhabitants crucified simply because they had resisted — then by the siege of Gaza.[6] Following that came the subjugation of Egypt, where the triumphant Alexander had himself crowned pharaoh.

The following summer Alexander recrossed the Euphrates River, where he found Darius waiting for him again with a massive force near the village of Gaugamela. Darius attacked and was repulsed. Then Alexander counterattacked with the light cavalry and Foot Companions, sweeping the field and sending the Persians reeling, a victory that effectively destroyed the Persian Empire. At the age of only twenty-five Alexander had conquered virtually every inch of the known world. But that conquest would prove insufficient.

Now he watches as his troops descend from the cold, windy mountain pass and begin pouring out into the warmer low country. He does not know where he is going, but then it really doesn't matter. He has almost no grasp of the geography beyond the rough outline provided years before by his teacher Aristotle, and while that has proven very rough indeed, it has hardly been an impediment.

Alexander is on the move once more, and for Alexander the prospect alone of further conquest is sufficient. As Stanton Coblentz writes, "If he was seduced by no thought of a quick western route to the East, he was allured by the kindred idea of reaching the end of the earth."[7] This is Alexander's adventure, and he will go anywhere riches and empire are rumored. But already he has heard the weary rumblings from within the ranks of his army.

The Greeks are now eight long years on campaign — an eight-year march away from Greece and home. Many do not understand where they are headed, or why they must push on. While Alexander's successes on the field have been unprecedented in the annals of martial conquest, those successes cannot quell the rumors of disloyalty, much less the growing murmurs of mutiny among his men. The grumbling grows more audible by the day. Alexander listens, but he does not hear.

Those eight years have not only exhausted his troops, they have also changed Alexander. His men want to go home. He can think only of pushing on, up and across the next high mountain or tumbling river for the mere chance to conquer still other worlds. According to Peter Green, "There was no predictable limit to his ambitions, only a constantly receding horizon *ad infinitum*. What he intended now was (in the most literal sense) a march to the world's end."[8] Alexander's soul has become infected with conquest lust, and his men just don't understand. They fail to grasp this one important thing: In his own mind Alexander is *aniketos*— invincible. He now believes he is a god.

What began in Macedonia as a genuine Homeric quest has descended into hubris and pretension of the most extraordinary variety. "We see a man who, inflated with the pomp of empire, insists on wearing the tiara and robes of Persian royalty; we view the poets and historians he takes with him on his expeditions in order to press-agent his mighty deeds; we hear his demand that his subjects, on approaching him, throw themselves to the ground in the Oriental fashion; we watch his self-veneration giving way to self-deification."[9]

In Alexander the newborn ego seems now to have lost all rational boundaries, to have inflated to the very edges of the universe, and it is that inflation that seems to be driving him. He rules now by whim alone, and often those whims are unnecessarily cruel. As Stanton Coblentz points out, Alexander has killed Cleitus, his closest friend, in a drunken rage; hanged Aristotle's nephew when he refused to prostrate himself; and executed the son of one of his generals, then the general himself in order to cover the dirty deed.[10] Indeed, there are also rumors making their way through camp that it was Alexander himself who had arranged his father's assassination. Wine, paranoia, and egomania are consuming him. Alexander seems to have devolved into a monster. Yet surely there is more to him than this.

For centuries it was fashionable to see Alexander as only the shining star, the youth extraordinaire exemplified in Greek mythology. Today it has become fashionable to see him as only the opposite. But the real Alexander cannot be reduced to either one or the other. He was, quite literally, both extremes, and that, I think, is at the heart of the enigma that surrounds him still.

In many ways Alexander *was* an extraordinary human being. Coblentz, hardly one of his admirers, readily admits that Alexander was "a man of great

Seven. Alexander and the Warrior Archetype

self-will and tremendous daring, a man of rose-clouded romanticism and wildly adventurous spirit, a man who was young and fiercely energetic, all-powerful and all-convinced of the ascendant star of his destiny."[11] Alexander was extraordinarily charismatic and absolutely fearless. In battle he had no peer.

It is *that* Alexander, a simmering cauldron of abilities, ambitions, and pathologies, who now descends into the Indus Valley, hoping to battle his way clear to the very ends of the earth. Along the banks of the Jhelum River he will soon fight the last of his great battles. In that engagement he will cleverly maneuver the opposing cavalry out of position, thus allowing his famed phalanx to move against the center of the Indian line of battle, which is a concentration of over 20,000 infantry more than four miles in length supported by some 130 battle elephants at intervals of approximately 100 feet.[12] Only the iron discipline and fierce training of the Macedonians will allow them to advance into the teeth of such a position without breaking, against a force of battle elephants, the mere appearance of which on other fields has often been known to demoralize and scatter opposing troops.

Alexander, leading the Companion Cavalry, begins the contest. Once the Indian cavalry has been dispatched with, he signals the phalanx to advance on the Indian center. The two forces collide in a contest of almost unimaginable horror, the Greeks hacking away at the crazed elephants even as they themselves are gored and stomped and mashed to pieces by the frightened beasts. Eventually the elephants give out, slowly backing away from the slashing Greek blades and spears, stampeding over their own horrified troops. The center of the Indian position collapses, and the Macedonians, in an orgy of bloodletting, hack the remnants of the Indian army to pieces.[13] The route to infinite conquest has now been opened for Alexander from the Jhelum River all the way to the Himalaya Mountains.

Alexander is intoxicated with the notion of further conquest, but his hoplites have had enough. The bloody, brutal, freakish assault against the raging elephants was for them the last straw, and they refuse to budge. Beaten, bandaged, and exhausted beyond measure, they want only to return home to Greece. Alexander tries every gambit in his bag of tricks to try to get them to head south toward the ocean, but a few days' march reveals a wide plain and the towering Himalayas beyond. The mere thought of climbing through *those* mountains is simply too much. At the mere sight of the towering mountains the Greeks rebel. Alexander, relying on his legendary charisma, addresses his Macedonian warriors, but nothing, it seems, will get them to budge. Finally Coenus, one of the oldest and most trusted of the group, stands and solemnly addresses Alexander. "Sir," he is reputed to have said, "if there is one thing above all others a successful man should know, it is *when to stop*."[14]

Andre Castaigne's 1899 charcoal drawing of Alexander the Great addressing his officers before battle. Always leading from out front, wounded on numerous occasions, Alexander is considered to this day one of the finest general officers in the history of warfare (Library of Congress).

Seven. Alexander and the Warrior Archetype

So, it is done. The trip back is almost as fraught with danger and battle as the one that brought the Greeks out of Asia Minor and an extraordinary tale in its own right. But eventually what remains of his army claws its way back to Babylon, where Alexander proceeds to consolidate his power and rule with an iron fist. In that manner he would remain until his untimely death on June 10, 323 B.C., in all probability a death occasioned by poison at the hands of assassins.[15] The great Alexander is dead at the age of only 32. His empire will crumble in a matter of days.

Over the course of the intervening centuries Alexander's life and death have spawned a cottage industry of biographers, their tales often ranging from romantic adulation to outright contempt, the individual stances generally dependent upon the developmental level that spawned them. As Peter Green points out, immediately after his death Alexander was generally detested as a self-adulating butcher.[16] Only later, as conquest again came into vogue with the Romans, was Alexander's image resurrected and dusted off for general adulation. There it more or less remained until the last few hundred years when, with the emergence of widespread level four thought and morality, he once again fell into disfavor.

Our interest in Alexander is neither to admire nor to denounce him. Rather it is to understand his place in the war-scarred history of *Homo sapiens sapiens*, to try to learn something of the convulsions the immature ego can go through as it begins to emerge. To begin with, it is important to realize that Alexander was not just raised as a warrior king. He was raised in relative luxury, tutored by no less a luminary than Aristotle, and ultimately inherited from his father the most sophisticated war machine the world had ever seen. Stanton Coblentz points

Bust of Alexander the Great from an etching by Jacques Reich. In a brief thirteen-year reign, the youthful Alexander conquered more territory than any other military chieftain in history, spread Greek culture across the globe, and left a lasting impression on the emerging Western psyche (Library of Congress).

out that "no man was ever more deliberately trained than Alexander to be a war-chief.... [L]ike Hannibal, he was the son of an outstanding general; like Hannibal he was taught by his father in the field (he led the cavalry at Chaeronae beneath Philip's very gaze). But unlike Hannibal, he was educated to be a king as well as an army leader."[17]

Unlike many warrior kings who were essentially feeble characters, Alexander appears to have been a man of extraordinary abilities: handsome, remarkably intelligent, shrewd, decisive, powerful, and athletic. Yet those positive characteristics, having in the warrior king no natural check or inhibitor, were in the end the very engine that allowed his ego to expand beyond all sensible and productive boundaries. No one, in the end, was about to tell Alexander what to do, and Alexander would brook no criticism.

Success drove Alexander to egomania, and in that egomania he eventually drowned, fancying himself—of all things—a god. Alexander developed a pathologically bloated ego, and as we have already seen, "the death fear of the ego is lessened by the killing, the sacrifice, of the other; through the death of the other, one buys oneself free from the penalty of dying, of being killed."[18] In that sense it is certainly no surprise that Alexander excelled at killing.

Yet it is the aimless, endless, virtually insatiable appetite for glorification that seems in Alexander his most striking characteristic. As Peter Green points out, his aim was literally to march to the ends of the earth, and it is here, I think, that Alexander begins at least to make some sense. For as we saw in chapter five, we are all under a natural pressure to grow developmentally, and only when that growth is thwarted do elements of hate and violence arrive on the scene. But in an ego as bloated as Alexander's, this pressure would have been felt not as a healthy urge to grow but rather as an intense *threat* to the ego structure itself, thus something to be avoided altogether, pacified, or projected.

Alexander, trained as a conqueror, simply misinterpreted the pressure to grow as a need to conquer, and since his ego knew no bounds, neither did his bloody ambitions. He then attempted to pacify this need through campaigning while projecting his death fear onto others. Just as with our serial killer in chapter five, the more hollow victories he achieved the greater his sense of desperation grew, until marching to the ends of the earth made a sort of sad and frantic sense to him. Today, bloated egos like Alexander's still stride resolutely across our world, but they are generally found in places like Wall Street or Hollywood or in athletics and politics where their insatiable lust for self-glorification is fortunately no longer expressed through the bloody conquest of others, but rather through the almost mindless acquisition of wealth, accolades, and political power.

So what shall we finally make of Alexander? I think Peter Green says it best:

His gift for speed, improvisation, variety of strategy; his cool-headedness in a crisis, his ability to extract himself from most impossible situations; his mastery of terrain, his psychological ability to penetrate the enemy's intentions — all these qualities place him at the very head of the Great Captains of history....[19] He spent his life, with legendary success, in the pursuit of personal glory, Achillean *kleos;* and until very recent times this was regarded as a wholly laudable aim.[20]

He was arguably the finest general officer the world has ever known, while unquestionably an egomaniac and butcher of the first order. In that sense we might call him *the* star performer of the level three world. While scholars still debate his legacy, it is certainly no fiction to suggest that Alexander left his mark on us all, and that fact is what is of critical interest to our inquiry.

Indeed, it is almost impossible to understand historical or contemporary warfare and the concomitant warrior ethos without understanding the imprint left by Homer's epic tales given form and substance through Alexander's life and campaigns. In a very real sense Alexander brought those early Greek ideals to life, providing as he did an extraordinary model for generations to come and giving the Homeric warrior ethos almost life everlasting. His Companion Cavalry, for instance, drawn from the families of the Greek aristocracy, would serve as *the* model for mediaeval knights.[21] The knight's mediaeval codes of chivalry would in turn form the basis for many of our current notions regarding honor, conduct, and decorum, and in that sense much modern behavior can be traced directly back to Homer via Alexander.

In the West these Homeric codes (the urge to close, in particular) would leave their marks, not only on the psychology of the individual officer and soldier at arms, but also on theories of tactical maneuver, firepower, and even weapon design and development for the next 2,000 years.[22] They would affect far more even than that. It was almost as if Alexander had unwittingly hijacked the emerging self and, by the sheer force of his personality, molded it to his likeness.

Historians have correctly noted that it was Alexander's campaigns that ultimately spread Greek culture around the known world and established that culture as the cornerstone of Western civilization. What they fail to take into consideration is that in doing so Alexander also stamped not only the Homeric codes on the military archetype but also the image of the glorious, youthful, divine conqueror on the emerging Western psyche. That imprint would have long-lasting effects. Robert O'Connell, hardly a romantic, remains nonetheless impressed by the Macedonian's legacy: "Alexander was personally an extremely attractive figure, a virtual incarnation of the youthful dream of transcendence. For this reason he became the center of a romantic myth which, in a sense, endures today."[23]

Not only would Western civilization have Plato, Aristotle, and Socrates to ponder, it would also have the glistening, romanticized illusion of Alexander himself to deal with for centuries to come. Stanton Coblentz writes as follows: "From Pompey and Caesar through the Crusader captains to Alexander's nearest kinsman, Napoleon, that romantic and magnified vision of the Macedonian adventurer has never left the world's consciousness, a goad and a justification to war-makers and above all ego-maniacal aggressors."[24]

Alexander was the product of an enormously capable but immature ego, the self-inflated persona of early adolescence that the Swiss psychologist Jean Piaget discovered during the course of his research and that we have already catalogued in some detail. But if — as noted in the preceding chapter — level four consciousness is home not only to the extreme egocentrism of an Alexander at its earliest stages but also eventually to the capacity to take the place of the other and to imagine what might be as well, isn't it reasonable to presume that somewhere along the line those deeper, far more mature abilities would ultimately collide head on with the immature motivations of someone like Alexander?

There were, after all, many approximate contemporaries of Alexander's — Socrates, Mohammed, Siddhartha Gautama (the Buddha), Jesus of Nazareth, Lao Tzu — who offered infinitely more meaningful notions of life than Alexander's empty lust for conquest, and who in many cases have made far more lasting impressions. So isn't it conceivable that the more mature and meaningful aspects of this new consciousness would come into direct conflict with the less mature as they emerged?

The answer, of course, is yes. As we shall see, much of the religious, cultural, and military conflict over the last 2,000 years has been between the maturing level four consciousness and its less mature predecessors. Many of these conflicts still rock our world today. Indeed, and as pointed out above, it was just that deeper level four emergence that eventually found the accomplishments of Alexander quite distasteful. And if level four consciousness can actually take the place of the other, then it seems only a matter of time (and a matter of logic) before that same consciousness will suddenly grasp all who are "others" as equals. If we are all equals, then it only makes sense that we should be treated as such, and from that fundamental level four comprehension will naturally flow the first spark of something called democracy.

That spark would take quite some time to finally ignite. But by 1776 those democratic notions would burst into flaming reality in the hands of a thirty-three-year-old legislator and scholar from Virginia. Thomas Jefferson would write this: "We hold these truths to be self-evident, that all men are created equal; that they are endowed by their Creator with certain unalienable rights, that among these are Life, Liberty, and the pursuit of Happiness."[25]

Seven. Alexander and the Warrior Archetype

Those are the words, of course, of the American Declaration of Independence, a synthesis of Enlightenment thought in which Jefferson would even go so far as to denounce the long existing slave trade which his own family had long been a part of. That portion of his original draft was later removed by Congress so as not to upset numerous Southern delegates, but even so the document remained a powerful affirmation of personal rights.[26]

But here is the really interesting thing. How exactly could Jefferson insist that these truths were in fact "self-evident" when for the entire prior human experience they had not, in fact, been self-evident to hardly anyone at all? Alexander did not fight for them. Homer did not write about them. Indeed, beyond a few European Enlightenment thinkers during the seventeenth and eighteenth centuries, such as Locke, Montesquieu, and Voltaire, it seems these "self-evident" truths had hardly occurred to *anyone*.

The answer is they had suddenly *become* perfectly evident to the evolving level four consciousness, a mind that for the first time could take the place of the other, and could therefore appreciate all people as equals. No prior level could comprehend such a thing nor think on such a wide, encompassing scale, and thus there had been no previous declarations of independence during the long course of human affairs, no personal rights that someone, somewhere had previously considered self-evident. Certainly some few brave and intelligent souls had taken a stand for logic, fairness and reason long before the American Revolution, but they had been few and far between. The American Declaration of Independence is nothing if not a clear example of *mature* level four thought, proof positive that after centuries of slow, at times tortured nurturing, level four had at long last arrived as a major player on the world stage.

With pen alone Jefferson catapulted the world forward, and from that striking document would soon flow a world opposed to slavery; a violent Civil War on the North American continent; a world of sit-ins, of freedom marches, and, ultimately, of encounters such as Tiananmen Square.

At the time Thomas Jefferson penned this declaration he was, of course, focused entirely on King George and the British Empire, but in reality he had declared independence from much more than Great Britain. With the Declaration of Independence level four abruptly announced its refusal to be shackled to all those prior levels that still struggled to confine and dominate the human spirit. It declared its independence from the god kings who by whimsy alone had ruled the planet for centuries — not just from King George, but also from Darius, and Nero, and Caesar, and, yes, even Alexander.

The year 1776 marked the arrival of something entirely new — the first, rebellious notions of democratic government. It was a watershed event in the affairs of *Homo sapiens sapiens*, a moment devoted to individuals who, for the first time in the long course of human events, would *demand* their individual rights. Level four consciousness had at long last produced a mind that would insist upon *equality*, and *justice*, and *freedom*. The modern world, our world, was about to explode upon the scene. And explode it did.

Eight

A New Way Emerges

Legend has it that Ptolemy removed Alexander's body for burial from Babylon to Alexandria, where it remained on public display for a period of time. Afterward Alexander's remains simply disappeared, swallowed, as they say, by the shifting sands of time, and no trace of either his body or his tomb has ever been located. Only his legend remains.

Slowly level four consciousness continued to emerge, gradually leaking out, over the centuries, more regularly into the general population. More and more people assumed the ability to think, not only deeply but also critically, to think not only about other people's thoughts but also their own.

Predatory wars too numerous to count continued to rage, of course. Indeed, for centuries predatory, level three war remained the dominant cultural theme across Europe, the Mediterranean, the Near and Far East, and ultimately even the Americas. It is not necessary to catalogue the carnage to understand the culture. The historic record speaks volumes to the predatory nature of this human epoch. So we will fast-forward this inquiry to the year 1521 and turn our gaze toward Europe. In April of that year nobles from all across Germany, Spain, and France hastily convened on the Rhine River in the small town of Worms. The occasion was a special church diet called to deliberate the theological writings of a suspect German monk.[1] It was serious business.

Inside the packed hall the thirty-eight-year-old monk was seated before a gathering of princes and officials representing what authority remained of the Holy Roman Empire — to say the least, a most powerful and threatening jury. The monk had been accused by Emperor Charles V of writing numerous theological tracts and pamphlets that were in direct conflict with the teachings of the Catholic Church, which was potentially a capital offense. The monk's name was Martin Luther, and he had been asked to stand and either renounce or accept his own writings, all of which were displayed as evidence before him.

Inside the hall the atmosphere remained close, hot, and intense.[2] While few understood exactly what was taking place, no one doubted that a great deal was at stake. Most fancied this forum nothing more than a hearing over church doctrine, but in reality it was one of the first few intellectual clashes between level three and level four thought and culture, two thought structures, history teaches us, that over time will not take kindly to one another.

Luther had been granted safe passage to Worms for his hearing, and while he arrived safely, his safe departure was hardly guaranteed. He knew he had the support of the powerful Frederick of Saxony, whose position had so far prevented the church from directly seizing him, and that of much of the local German population; but he could hardly be sure what the church might decide, or who might or might not step forward to protect him in the future. Indeed, his writings had become so popular among the common people that the papal authorities in Rome eventually came to see him as a dangerous force that had to be eradicated. And Luther knew that, before the awesome power of the church, his popularity alone would guarantee him nothing.

Luther's principal attacks had been directed at the papacy and particularly at the "notion that God was expected to reward a Christian in proportion to the number of prayers said, of pilgrimages undertaken, of contributions made; the cult of saints and their relics, with its odor of polytheism; and the sale of indulgences."[3] Luther's pen had been scathing, unrelenting, and, worst of all from the church's perspective, logical. Above all else, it was the sensibleness of Martin Luther's arguments that carried the day. Luther made sense and the church did not, and in just that way *both* seemed in the end to be on trial. The only question was whether power or logic would prevail.

Martin Luther had been born in 1483 in the German town of Eisleben. His father had worked his way up from peasant to the mining business and gradually accumulated enough wealth to send Martin off to college in order to study law. But the law held no fascination for Martin, and in 1505 he gave up his studies and entered a monastery in Erfurt.[4] Extremely bright and well educated, it would not take Luther long before he began to question the wisdom of much that passed for sixteenth century Catholic religion.

Thus he came to face a jury that could well demand his life if he was found guilty. On a mid–April day he had but two choices: to either stand by his writings or renounce everything he had come to believe. The latter was simply unthinkable. So Martin Luther stood, reaffirmed the works as his own, then told the assembled noblemen that unless convicted by either scripture or reason, he would recant not a word. He believed, he explained, not in the dictates of popes or church councils, but only in the power of God as revealed through scripture, and in the mind as it had the power to reason. Standing alone before the enormous power of level three state-sanctioned religion, his last words to the assembled nobles were reported to have been these: "Here I stand, I can do no other."[5]

Level four consciousness is here on display — curious, logical, scientific in its thought processes, comprehensive, stubborn, fearless. Luther could no more denounce those things he had come to know as true than a leopard could change its spots. Nurtured for centuries by the early kings and their courts, level four had appeared in the general population at long last, this new rational mind, and it would not be long before it began to change the world.

Yet, because access to the most favorable developmental *conditions* for a long while remained very limited, this change had come slowly and, initially at least, to but a very few. Only 5 percent of the German population could even read at the time of Luther's trial, and Martin Luther was one of the very few of them who would be favored with a college education.[6] It would be 1600 years after Alexander's death, for instance, before Europe would stir to life with the great revival of art, literature, and learning known as the Renaissance. And it would not be until the year 1564 that Galileo Galilei, the man considered by many to be the founder of modern science, was born.

While those developments would prove historic, it is important to bear in mind that, as Paul Tillich points out, the Renaissance was the collaborative effort of perhaps no more than a thousand people, and Galileo was promptly charged, tried, and convicted by the church's Inquisition for publishing his astronomical findings. Even by the late sixteenth century it is clear that level four consciousness had established but the smallest beachhead in the general population; and for those positively affected, the long arm of level three mythological thought remained happy to snatch them off the street and kill, torture, or imprison them for the heretical thoughts and facts they espoused.

Yet reason would prove such an extraordinary tool for exploring the natural world that, once unleashed, like a wildfire it would burn away many of the prevailing mythic dogmas with nothing more than the searing heat of its intense gaze. Reason provided a whole new way of looking at the world, and the results it uncovered could not easily be ignored. Ken Wilber says "slowly, in both the East and West, began to emerge rational philosophies, rational sciences, rational policies, rational religions — some of which indeed pointed *beyond* reason, but all of which depended on reason as a platform that would secure a common and mutual understanding for anyone, of any color, race, or creed."[7]

The slow rate of this emergence was a result of the lack of the requisite developmental conditions needed to support level four growth in the general population and the sheer depth and scope this developmental level itself offered. Indeed, Jean Piaget considered level four to be so substantial in its dimensions that he divided it into two substages, the first rather crude and incomplete, the later substage demonstrating the final emergence of true level four capabilities.[8] In that sense the world would not become fair, logical, and just overnight. Indeed, even today for most people across the globe — a globe still ruled mainly by level three thought and government — freedom and justice remain little more than fond aspirations. So, those of us fortunate enough to reside in a democracy might consider our lot a fortunate one, for, as we shall see, democracy is the natural governmental expression of the level four thought process.

Martin Luther's efforts would drive a wedge through the center of the Catholic Church and in so doing inaugurate the Protestant Reformation. Convicted by the pope's court at Worms, Luther was nevertheless spirited away by friendly forces to live out his days in safety in Germany. While his Reformation would reshape the Christian church, many of the new, reformed, Protestant churchs would soon settle into the same mythic ruts from which they had sprung. The Reformation was happy to call into question the excesses and doctrinal confusions of the Catholic Church, but it would fail, by and large, to examine its own. And while Martin Luther surely represented the new breath of reason (the first substage of level four), it was a reason less than entirely rational and scientific. In *Psychohistory and Religion*, for instance, Paul Pruyser points out that Luther "continued to believe firmly in the concrete reality of the devil, advocated the burning and torturing of witches, recommended the drowning of idiot children, and retained a considerable element of magical thinking in his approach to the sacraments."[9] This is neither shocking nor unusual. As demonstrated, new stages emerge gradually, and often the remnants of lower stages remain in ready evidence. But Martin Luther's stand against the Catholic Church — like the trial of Socrates more than a thousand years before — would signify one of the first great clashes between level three and level four thought. Many, many more would follow.

By the late sixteenth century and on into the seventeenth, for instance, the level four mind would truly begin to flower, and so profoundly sweeping was the thought process inaugurated during this period that it is no surprise that this era (at least in the West) has been called the Age of Reason, or simply the Enlightenment. For society in general it represented a breathtaking transformation. "Time-honored certainties crumbled: Old assumptions about the authority of kings, the structure of the universe, even the very existence of God, were called into question. The thinkers of the age would take nothing on trust; old habits of unquestioning obedience to religious, political, and social authority were replaced by the scrutiny of all ideas under the penetrating light of human reason."[10]

In 1642 — indeed, on the very day Galileo passed away — Isaac Newton was born in England, and the struggling, emerging science that Galileo had first given life Newton would take in hand and reshape into a mighty force. The impact of his reasoning was immeasurable. Newton "proved through painstaking observation and scholarship that the solar system was governed by the laws of gravitation. His discovery opened a door to the mysteries of the universe, offering a simple — but rigorously tested — mechanical principle that could be applied to the movements of celestial and earthly bodies alike."[11] That same level four science would have a profound influence on the world and, naturally enough, a profound influence on war. Two of these developments would eventually begin working against the Homeric archetype and the culture of predatory warfare it had long sustained.

First, the new scientific method — Galileo's and Newton's science of testing and observation — would in only a few centuries produce a widespread Industrial Revolution that would not only reshape the implements of war but by its very nature also inaugurate an arms race that would quickly spin completely out of control. It would not be long before scientific technology would go from producing weapons that armies could effectively use in the field (rifles, mines, rockets, machine guns, tanks, etc.), to producing weapons that armies dared not use at all (atomic and hydrogen ordnance). In a matter of less than one hundred years (1860–1960), war would be transformed from a heroic undertaking into a doomsday proposition.

Second, as level four consciousness emerged in the general population it would, as pointed out in the previous chapter, inaugurate the first democratic notions. Historically this phenomenon had occurred before very briefly in places like Greece and Rome, but those democratic, republican notions had very rapidly been crushed under the sheer weight of level three predatory opposition. Not so this time around. Enlightenment thinkers such as Voltaire, Locke, and Montesquieu would forge the intellectual framework for these radical ideas, while the British colonies in North America would prove the perfect, albeit unintended, incubator for their growth.

Those ideas, matured by centuries of tinkering, would eventually collide headfirst with the old mythic establishment. The most celebrated of these collisions took place between colonial minutemen and British regulars in 1775 at Concord Bridge in Massachusetts, an incident so fateful it has ever since been called the "shot heard round the world." That encounter would lead directly to the American Revolution, and that event to a sweeping, worldwide revolution of level four thought that remains ongoing to this day.

Something new, something radical had begun. As Stanton Coblentz puts it, "The English looked upon the colonists as subjects, who owed them duties politically and economically; the Americans, reared in the traditions of freedom and accustomed to self-reliance in their rugged frontier environment, were impatient even of the slightest restraint."[12] Thomas Jefferson's Declaration of Independence would set that "shot" (the democratic process) into motion, and that motion would be given form in June 1788 with the final ratification of the Constitution of the United States of America. Very suddenly, radical thought had been given radical structure, and the entire world has been dealing with the upheaval ever since.

Democracy, the form of government created by the level four mind, tends to place ultimate power in the hands of the people, and additionally to enforce a separation of powers within the structure of government itself. These two developments would severely limit the power and authority of the head administrative official (the old warrior king), thus severely limiting the ability to make war without the backing of the body politic. In a very real sense democracy made the warrior king obsolete overnight. Democratically elected governments would still fight wars, of course, but from now on they could be inaugurated and fought only with popular support and for reasons the public accepted as worthwhile, a severe counterweight to the sometimes fanatical tendencies of the warrior class.

Because as it matures level four develops the ability to take the place of the "other," democratic populations in general do not hunger for war. The xenophobic tendencies of level three aggression gradually give way to a far more tolerant and inclusive state of mind that finds war, except for the most serious of causes, an unacceptable proposition. While some few predatory wars have been initiated by democratic governments (the Mexican-American War of 1847 by the United States, as an example), most wars have not been inaugurated by democratic nations but have in fact been responses by those nations to predatory, level three aggression. World Wars I and II serve as obvious examples of the latter, and both will be dealt with in coming chapters as this inquiry continues.

Unfortunately, level four passiveness and willingness to compromise are very often misinterpreted by level three minds as weakness, and level three, wedded for centuries to the predations of military power, has only contempt for weakness. This is the reason dictatorships always seem to conceive democracies as weak and corrupt: level three simply cannot grasp democratic principles, and is generally confused, even appalled by the open display of common freedoms — and level three minds often radically miscalculate their moves as a result.

Once aroused to war, however, level four nations can become formidable opponents, generally displaying a fortitude coupled with organizational, technological, strategic, and tactical capabilities level three never dreamed possible. The surprise Japanese attack on Pearl Harbor that forced the United States into World War II serves as a classic example of this. While from the Japanese perspective the Pearl Harbor attack was without question a tactical masterpiece, from an overall strategic point of view it would prove little more than a catastrophic miscalculation. And ultimately it would be the death warrant of the Japanese Empire.

An even more striking example of the same phenomenon is the now almost forgotten Winter War of 1939 fought between the Soviet Union and its tiny neighbor, Finland. Joseph Stalin, after signing a nonaggression pact with Adolf Hitler, was quick to chew up many of the small states along his borders such as Latvia and Lithuania. Next up for immediate conquest was Finland, a small republic sitting just north of Estonia on the Baltic Sea.

The Soviet army was massive and modern. The Finnish army was paltry and ill-equipped. On paper a conflict between the two appeared to be no contest, and Stalin expected to swallow Finland whole in a matter of weeks only. The Soviet invasion commenced in November, but it immediately bogged down against what can only be described as furious Finnish resistance. The Finns — fighting in the snow and cold with stealth, intelligence, and enormous spirit — staggered an invading force five times their own size. Ultimately, while giving ground, they fought the Soviets to a standstill throughout that long, cold winter, inflicting catastrophic casualties on the Russians while preserving both their freedom and independence.

To put it mildly, level three and four neither understand nor appreciate one another and much of the tumult across the globe over the past three centuries — from the Inquisition to the American Civil War; from the Scopes Monkey Trial to September 11— can be directly attributed to the volatile clash along this psychological fault line. Level three tends to be imperial, agrarian in its culture, religiously fundamental, and socially stagnant, while level four tends to be passive or even isolationist in its outlook, industrial, religiously tolerant, and socially upwardly mobile. These levels are, to say the least, almost virtual opposites.

The impact of level four emergence during the seventeenth and eighteenth centuries was demonstrated across the span of human culture exactly as might be expected, and it was principally reflected in two distinct motivations. The first, driven by the new egalitarian understanding, quite naturally was the demand for individual rights and freedoms. Obvious examples of this are the American and French revolutions, along with the political push ultimately for those same freedoms in other countries such as Great Britain.

Just as early level four includes the emergence of the formative ego for the individual, this period also involved the first widespread demonstration of the new collective ego: the modern nation-state. Coalescing out of a galaxy of junior states, regions, and principalities, rough new nations emerged from a backdrop of disharmony. Germany, Italy, France, and Spain serve as examples of this movement toward consolidation. Indeed, in many areas across the globe, this process is still ongoing, and the attendant conflict is obvious. Thus this period was, as the world remains, rife with wars for both national consolidation and, simultaneously, open rebellions against the old, level three monarchial (tyrannical) structure. Level three would resist this tidal wave of change with every fiber of its being, and this stubborn resistance would create a period of enormous tension, chaos and, to say the least, upheaval. Robert O'Connell says, "The ancien régime was a bulwark against change in a time of transition, a delicately balanced mechanism poised on a volcano. Given the forces to which it was subject, it proved remarkably tenacious."[13] Tenacious is to say the least. In many regions around the world today, this resistance is active still, and level four democratic notions have yet to make an inroad. The Middle East serves as a prime example.

Revolutionary wars predicated upon the demand for human rights altered the face of warfare. These were wars driven now by ideas, not greed, booty, or simply a taste for limitless expansion as Alexander had known. Just as with Martin Luther's theological writings, imperial level three empires would find it very difficult to even understand, much less quash, the extraordinary power of ideas with the force of arms alone. The old tactics previously employed in warfare seemed for some odd reason no longer applicable, for democratic ideas appeared to have a life all their own. British military frustration over the lack of success during the American Revolution exemplifies this point. Cities were taken and occupied, large territories seemingly pacified, revolutionary armies vanquished, but still the war dragged on. Increasingly it seemed impossible to deliver a decisive blow. All the time-honored rules of European warfare appeared frustrated. British general Charles O'Hara detailed his frustrations in a letter home: "It is a fact beyond doubt that their own numbers are not materially reduced, for in all our Victories, where we are said to have cut them to pieces, they very wisely never staid long enough to expose themselves to those desperate extremities.... [H]ow impossible must it prove to conquer a Country, where repeated success cannot ensure permanent advantages, and the most trifling check to our Arms acts like Electric Fire, by rousing at the same moment every Man upon the vast Continent to persevere upon the most distant dawn of hope."[14]

As pointed out above, this would also be a time of rapid technological evolution in weaponry, most principally the period during which firearms evolved to the point that they would take command of the battlefield. Firearms originally appeared around 1300 but would not be developed into capable battlefield ordnance until the late fifteenth century. Initially this would not greatly destabilize the general symmetry of weapons that existed between Western combatants, but with the coming of the Industrial Revolution, all that would change. Soon the killing capacity of battlefield ordnance would vastly increase, turning the sort of encounters that only a few decades previous had been relatively bloodless into spectacles of slaughter.[15]

Firearms also posed a direct threat to the old Homeric codes of heroic warfare, and the reason for this was simple. In a relatively short period of time any peasant could be trained to load and fire a gun, thus greatly devaluing the need for training, courage, and martial excellence. To approach for close combat an adversary armed with a gun was no longer heroic, but suicidal. A well timed volley from a line of infantry or battery of artillery could wipe out the most trained and courageous group of warriors long before they ever had the chance to close with the enemy. The finest officers and men could be killed from a distance by weapons that might only be heard, and by an enemy that often could not even be seen. Thus the Homeric codes began to wither and die under frightful volleys of small arms ordnance.

Yet this obvious fact would go virtually ignored by military planners for almost three hundred years. The preferred method of combat would remain the infantry frontal assault, the bayonet advance, or the cavalry charge undertaken by massed ranks into the very teeth of a fixed position, seemingly regardless of the human cost. Like a ghost, Homer's shadow still hung over even the nineteenth and twentieth century battlefields and, coupled with the increasing lethality of weapons, the toll in lives and misery would prove catastrophic.

During the eighteenth century echoes of Alexander's predatory ego would arise in the likes of Louis XIV of France, Charles XII of Sweden, Peter the Great of Russia, Frederick II of Prussia—all admirers of Alexander—and finally Napoleon, who literally attempted to emulate the Macedonian king. In both America and France level four revolutions erupted, rattling the world with novel notions of freedom and democratic government.

The modern map of the world began to coalesce in the nineteenth century out of a backdrop of shifting boundaries. As Stanton Coblentz points out, "the Crimean War, the American Civil War, the Franco Prussian-War, the Boer War, the Spanish-American War, the Russo-Japanese War, the Balkan Wars"[16] were all products of this new sense of national self-determination driven by the arrival of the collective level four ego. Many of these new nations would, initially at least, maintain the old level three drive to empire, but over time that drive would subside.

Yet all of these events were preceded by the American Revolution, an event, as previously described, that proved enormously important in the emergence of level four thought, law, and government. That revolution would be unique in many ways, a war driven by *ideas* not conquest, those ideas themselves redefining the relationship of the individual to government and as a result the course of human events. It was a revolution that changed not only the shape of war, but the shape of history. To better understand the evolution of warfare it is essential to grasp the genesis of revolution.

Nine

Revolution

The American Revolution was one of the most unique and important occurrences in the long and often troubled affairs of humankind. Indeed, it represents the first significant, enduring burst of level four consciousness that, up until that point, the world had ever experienced. But in terms of the emergence of revolutionary ideas, while it might have been the first burst, it would not be the last; thus an examination of its origin and outcome is surely in order.

A sensible case can be made that the Revolutionary War in America had its roots in the successful conclusion of the Seven Years' War fought between Britain and France; this, on the North American continent, was called the French and Indian War. This case postulates that when the French and Indian War ended in 1763 Britain, although the victor, had been utterly drained of resources and, seeking to reestablish financial stability, initiated a series of taxes on her North American colonies in order to help foot the bill. In that Britain maintained garrisons in the colonies, and particularly in the western territories to help shield the colonists from the threat of repeated Indian attack, the British government considered these taxes to be both reasonable and equitable.

The colonists, on the other hand, rankled by what they perceived as unfair and excessive taxation, became increasingly hostile to both the taxes and the government that had imposed them. The British government refused to compromise and the situation devolved eventually into open hostilities. Those hostilities led ultimately to war outright. But to suggest that an entirely new nation, conceived on the rudimentary elements of democratic government and individual freedoms — something the world had never before witnessed, had sprung to life due to a mere tiff over taxes, is to suggest the implausible. The taxes put in place by the royal government in London beginning in 1764 may well have provided the spark that ignited the fire, but the fuel the fire required to burn for years and years ran far deeper than issues of taxation. Indeed, these issues had been accumulating slowly to the task for 150 years,

Nine. Revolution

what would become the United States slowly evolving into form over that same period of time.

To understand this further it might do well to ask a simple question. When, exactly, did this thing we today call the United States of America actually come into existence? The quick answer might be that the United States came into existence with the formal adoption of its constitution during the late summer of 1788,[1] but that answer ignores the obvious fact that the nation had been functioning since the adoption of the Articles of Confederation in 1781. So then it might be argued that the United States began with the Articles of Confederation. But that would fail to take into account the fact that the United States had operated for almost six years of conflict prior to the confederation government under the Continental Congress, and it also fails to take into consideration the fact that the Declaration of Independence had been issued on July 4, 1776, long before either the Articles of Confederation or the Constitution existed. So it would seem to make sense to date the birth of the United States to July 4, 1776, except for the fact that a Continental Congress had already met much earlier, in 1774, and that by the summer of 1775 a continental army—known by various names—had already fought a pitched battle with the British at Breed's Hill (known as the battle of Bunker Hill) and remained in open conflict with the British over Boston. Indeed, the Continental Congress had already dispatched George Washington to Boston to take command of this army, such that it was, almost a year before declaring independence. Thus the United States, as a political reality, seemed at that point already to be in some form of quasi-existence.

The sequence of events at Boston that had touched off the military confrontation between the colonists and the British began essentially with the Boston Tea Party in 1773 when angry colonists tossed British tea into Boston Harbor rather than pay the required export duties. This led to a series of punitive measures taken by the British government known as the Coercive Acts, essentially shutting down the port of Boston and crippling the Massachusetts Assembly. Additional British troops were sent to Boston to enforce these measures, and when those troops marched off to confiscate stores of arms supposedly maintained at Concord they were met by minutemen on the green at Lexington and later at Concord Bridge, and gunfire was exchanged. American militia then chased the British soldiers back to Boston, inflicting severe casualties on the marching column, and soon a regional colonial army, responding to calls of alarm, descended upon the city.

David McCullough explains the situation:

> In April, when the call for help first went out after Lexington and Concord, militia and volunteer troops from the other New England colonies had come by the thousands to join forces with the Massachusetts regiments —1,500 Rhode

> Islanders led by Nathanael Greene, 5,000 from Connecticut under the command of Israel Putnam. John Stark's New Hampshire regiment of 1,000 had marched in snow and rain, "wet and sloppy," "through mud and mire," without food or tents, seventy-five miles in three and a half days. The Massachusetts regiments, by far the strongest of the provincial troops, possibly numbered more than 10,000.[2]

The colonials at Boston numbered perhaps 16,000[3] in total and probably outnumbered the British two to one.

The colonial force had seemingly risen from the soil, and few at the time knew even what to call this ragtag accumulation of men.

> Even now it had no flag or uniforms. Though in some official documents it had been referred to as the Continental Army, there was no clear agreement on what it should be called in actual practice. At first it was referred to as the New England army, or the army at Boston. The Continental Congress had appointed George Washington to lead "the army of the United Colonies," but in correspondence with the general, the President of Congress, John Hancock, referred to it only as "the troops under your command." Washington, in his formal orders, called them the "Troops of the United Provinces of North America."[4]

The point of all this, put simply, is that it is impossible to pin down to any degree of certainty just when the United States actually came into existence, the reason being that the final reality of the United States was not so much an event as it was an *evolution* of events. And because the American Revolution was as much a matter of thought as it was bullets, it is also difficult to define precisely when that revolution began and when it ended, if, indeed, it has ever ended. While arbitrary dates can certainly be ascribed to both the beginning and the end, the reality of it seems to be that the revolution certainly began long before July 4, 1776, and ended long after the defeat of Cornwallis at Yorktown in 1781. The reason for this is that the American Revolution was the result of the slow maturation of level four consciousness on the North American continent, and thus a process that had been long in the making.

As discussed earlier, the psychiatrist Clifford Anderson noted of maturational growth that developmental advances "reflect movement down a preexisting path — a template every bit as inherent as the physical one that creates the brain, the heart, and the muscles in the arm."[5] In other words, level three will naturally advance to level four if the proper *conditions* are in place, and what we find on the North American continent as the first settlers arrived were the very rudimentary but basic conditions for developmental advance. What were they?

It goes without saying that the settlers who established the first permanent outposts — those who arrived at Jamestown in 1607 or Plymouth in 1620 — faced a harsh and uninviting landscape upon arrival. While these settlers were

The Battle of Bunker Hill, by E. Percy Moran. Actually fought on Breed's Hill outside of Boston on June 17, 1775, before either the Declaration of Independence had been issued or George Washington had been appointed commander of the Continental troops. The battle would prove a technical British victory, but a victory gained at such a terrible price that the outcome greatly inspired American confidence and resolve (Library of Congress).

often highly motivated, and some reasonably literate and accomplished, they were generally unprepared for the wilderness they would be called upon to master and often incapable of food production, even the construction of basic housing. For those who survived, however, the New World offered a land of relative peace, prosperity, and freedom that Europe — for the middle and lower classes, at least — could not come close to matching. "Life for the earliest settlers had been a grim struggle against disease, hardship, and Indians, who were resentful and fearful of the land-hungry intruders. However, with each new generation prosperity had increased, and with it the population. In 1688, just 200,000 people lived in the British-owned colonies, but by 1750, that figure had risen to more than 1.5 million."[6]

Britain looked upon her colonies with essentially a commercial eye, and with London three thousand miles away, the colonists were left by and large to run their everyday affairs without much in the way of royal interruption. Each town soon had a council that deliberated local ordinances and adjudicated local laws, while each colony had some sort of assembly in the image of the English Parliament. This system, small and primitive as it was, provided the colonies with at least 100 years of democratic tinkering and development prior to 1776, and virtually all of the delegates who arrived for the Continental Congress had prior legislative experience as a result.

As the population grew, institutions of higher learning slowly emerged. In 1636, for instance, Harvard College was founded outside of Boston, followed in turn by the College of William and Mary in 1693 and Yale in 1701. Others would soon follow, and education became enshrined as one of the principle American means to upward mobility. While a college education was generally for the more well-to-do, the colonies nevertheless began to produce their own class of doctors, engineers, ministers, etc., and were no longer reliant upon Great Britain for the sum total of their professional expertise. The colonies were becoming self-sufficient.

Most important, this rising class of educated professionals appears to have been one of the first worldwide to emerge without a strong association with the ruling military elite, which was the maturational template that had existed since the dawn of civilization, and thus a fundamental grounding in warfare as a means to both status and career. Indeed, in the colonies there was no military elite per se, and this educational variance would have a profound effect upon how this rising class envisioned both themselves and the world in which they moved. Schooled in the classics, yet awash in Enlightenment thinking, these young political practitioners and philosophers would craft a new form of government, a governmental format no longer predicated exclusively upon power but upon liberty and civil rights. In that sense the North American colonies represented a sort of primitive incubator for democratic notions, not merely in theory, but also in practical applications at both the community and regional levels. It would just be a matter of time before that experience would blossom into something larger and far more encompassing.

This incubator evolved for many reasons and over many years, and the conditions upon which it was founded were those that eventually led to level four growth on an unprecedented scale. In the New World there was an abundance of land upon which to settle and over time a reasonably stable and healthy diet that nurtured both body and mind. While life in rustic North America was certainly harsh, it was at least free of the endless wars and hatreds that had wracked Europe since the days of Rome. Thus it allowed individuals to live and grow in an atmosphere of relative tranquility. Forced to master a

Nine. Revolution

hostile landscape thousands of miles from ruling London, the settlers became remarkably self-reliant, a situation that bred self-confidence and pride coupled with independence of thought and action.

These, then, were the farmers, trappers, merchants, shopkeepers, and others who peopled that unnamed army that laid siege to Boston in 1775. Dirty, untrained and profoundly undisciplined, they were highly motivated but knew almost nothing of war and were apparently not much to either smell or look at. Indeed, the aristocratic British officers unfamiliar with the fighting capabilities demonstrated by the Americans during the French and Indian War often mistook them for the worst sort of riffraff and rabble. "To the British and those Loyalists who had taken refuge in Boston, they were simply 'the rebels,' or 'the country people,' undeserving the words 'American' or 'army.' General John Burgoyne disdainfully dubbed them 'a preposterous parade,' a 'rabble in arms.'"[7]

The British, unschooled in the spontaneous assembly of a free people, would of course mischaracterize the scene before them, and for this they can be forgiven. The American army was, after all, a phenomenon of sorts, and certainly a phenomenon never before experienced by the British ruling class. The army was initially composed of a wide cross section of men from across New England, of all ages and occupations, some as young as ten.[8] David McCullough tells the story of young John Greenwood, who had hiked 150 miles on his own to join the American cause.

> John Greenwood, a fifer — one of the more than 500 fifers and drummers in the army — was sixteen, but small for his age and looked younger. Born and raised in Boston, he had grown up with "the troubles" always close to home.... Thrilled by the sound of the fifes and drums of the regulars occupying the city, John had somehow acquired "an old split fife," upon which, after puttying up the crack, he learned to play several tunes before being sent to live with an uncle in Falmouth (Portland), Maine. In May 1775, hearing the news of Lexington and Concord, he had set off on foot with little more than the clothes on his back, his fife protruding from a front pocket. All alone he walked to Boston, 150 miles through what was still, much of the route, uninhabited wilderness. Stopping at wayside taverns, where troops were gathered, he would bring out the fife and play "a tune or two," as he would later recall. "They used to ask me where I came from and where I was going to, and when I told them I was going to fight for my country they were astonished such a little boy, and alone, should have such courage."[9]

Courage abounded and faith abounded, the ready belief that men imbued with the spirit of liberty alone would naturally prevail over mere professional soldiers with no true cause to speak of. General Washington, on the other hand, one of the few colonial officers with actual military experience, held a contrary view. Washington was a realist, and he was initially discouraged by the army he found waiting for him near Boston. Joseph Ellis, a biographer of Washington, explains:

> Based on his earlier experience as commander of the Virginia Regiment, reinforced by what he witnessed on a day-to-day basis at his Cambridge encampment, he became convinced that an army of short-term volunteers, no matter how dedicated to the cause, could not win the war. "To expect then the same service from Raw, and undisciplined Recruits as from Veteran Soldiers," he explained, "is to expect what never did, and perhaps never will happen." His convictions on this score only deepened and hardened over the years, but from the start he believed that militia were only peripheral supplements to the hard core, which needed to be a professional army of disciplined troops who, like him, signed on for the duration.[10]

Over time, the militia would give way to hardened, veteran soldiers, and eventually confidence in the cause alone would morph into a more realistic understanding of the realities of war; but the spirit of liberty never ceased to motivate the Continental Army and militia units always played an important role in the revolution. It is interesting that 16-year-old John Greenwood had left to fight, not for his state or region but for his "country," long before the United States had a name or formal identity. In that sense the country seemed already to have existed in some way and at some level in the minds of those who marched off to fight the revolution, its future form and independence the only real issues. That the United States did not yet have a name or a flag or a date of birth, does not mean that it did not have both shape and meaning long before militia units confronted British regulars at Lexington and Concord and confirms again the notion that the birth of the United States was not an event but an evolutionary sequence that had been long unfolding.

The Revolutionary War itself, fought over an extended span of more than six years, in retrospect seems more an exercise in the art of persistence than it does the art of war. Indeed, despite numerous blunders, and equally numerous startling triumphs, the colonial forces won in the end by simply refusing to lose as much as they did by the employment of any exceptional tactical schemes or arrangements. By and large this could be attributed to their leader, General George Washington, who led the continental forces from beginning to end and as much as anything else simply refused to be beaten. Never having commanded a force larger than a regiment prior to the war, he nevertheless learned from his mistakes and by war's end commanded an international reputation virtually unequalled. As Joseph Ellis notes,

> He was not, by any standard, a military genius. He lost more battles than he won; indeed, he lost more battles than any victorious general in modern history. Moreover, his defeats were frequently a function of his own overconfident and aggressive personality, especially during the early stages of the war, when he

escaped to fight another day only because the British generals opposing him seemed choked with the kind of caution that given his resources, Washington should have adopted as his own strategy. But in addition to being fortunate in his adversaries, he was blessed with personal qualities that counted most in a protracted war. He was composed, indefatigable, and able to learn from his mistakes.[11]

It can be said that the war began at Lexington in 1775 and ended at Yorktown some six years later, but in between there were any number of interesting events from a strategic, tactical, and developmental perspective. Washington began the war it seems, as Joseph Ellis points out, determined to beat the British at their own game; to inflict a smashing defeat upon the British army that would both win independence for the colonies and simultaneously verify Washington's own skill as a commander at war. But that desire rather quickly ran afoul of reality. Washington was able to force the British to abandon Boston in 1776 by cleverly fortifying Dorchester Heights — a tactic urged upon him by his staff — which commanded the harbor, but hope of a quick war and speedy recognition of the new nation soon unraveled at New York.

Hoping to fend off a new British force of far greater magnitude than the one that had abandoned Boston, Washington foolishly decided to defend New York City from attack, in essence placing his own army in exactly the same precarious posture the British had been in at Boston. The obvious flaw in Washington's defensive scheme was that waterborne access to New York for the British fleet was almost unlimited, thus a sizeable force could be with ease landed *behind* colonial lines, cutting off Washington's entire army from escape at any number of points. After a series of near catastrophic blunders — and one remarkable nighttime evacuation of Brooklyn — the beaten Continental Army was forced to abandon New York and begin a demoralizing trek across New Jersey south toward Pennsylvania and the distant hope of safety. In New York Washington had enjoyed an army that numbered nearly 20,000 men. After his harassed retreat across New Jersey his force numbered less than 3,500.[12] It seemed the end of the brief revolution was near.

The Continentals crossed the Delaware River at Trenton, and Washington wisely had every boat within sixty miles either taken or destroyed so that the British could not easily pursue. It was a prudent move that would have far-reaching consequences. The British general, William Howe, upon reaching the Delaware fancied the colonials essentially whipped, and he suspended further campaigning until the arrival of the coming spring.[13] Howe then shifted the bulk of his army north into winter quarters, leaving a string of small outposts across New Jersey to command the ground taken while pursuing Washington to the banks of the Delaware. One of these outposts was at Trenton, just across the Delaware from Washington's position at Bristol, Pennsylvania, where Howe had garrisoned a detachment of mercenary Hessian troops. It was from Bristol that Washington and his defeated army rebounded, launching a small but dramatic attack on Christmas night 1776, an attack that would save the revolution and a victory that would shock the world.

Through snow and ice Washington led his small command on one of the most audacious and desperate ventures in American military history, leading approximately 2,400[14] men on a surprise attack against the Hessian garrison at Trenton that numbered somewhere between 2,000 and 3,000 effective troops.[15] For the cause of American independence this assault was surely do or die. Fittingly, the password for the operation was "Victory or Death."[16] Frozen and exhausted — two of the colonials froze to death during the march — Washington launched his attack in the early morning hours and routed the Hessians in less than an hour, capturing most of the command. That night Washington's victorious and exhausted army slipped back across the Delaware to safety.

It was genius born of utter desperation, and in the blink of an eye George Washington had flipped the calculus of victory and defeat on its head. While the victory at Trenton was minor in strictly military terms, its psychological effect across the beleaguered colonies was momentous. Then, a week later, Washington struck again, once more slipping across the Delaware and moving back to the vicinity of Trenton. After Washington's Christmas victory, General Howe had dispatched General Charles Cornwallis with a sizeable force to bag Washington once and for all. Cornwallis soon located Washington's troops near Trenton and launched his initial assault as light faded into evening. Thinking he had what remained of the Continental Army cornered, Cornwallis waited for dawn to renew his attack, only to discover at first light that Washington had slipped away. Where had he gone? Cornwallis had no idea.

In fact, Washington had marched via back roads — with campfires left burning and cannon wheels wrapped to muffle any sound — on a wide arch away from Cornwallis and attacked the unsuspecting British rearguard at Princeton, there overwhelming the British in a savage contest on the outskirts of town. This colonial victory proved yet another stunning defeat for the

Opposite, top: Washington leads his troops into Trenton, Christmas Day 1776, culminating one of the most audacious and desperate missions in American military history. The print is from an engraving created by the Illman brothers, 1870. *Bottom: Washington at the Battle of Princeton.* Riding between opposing lines of battle, Washington rallied his Continental troops, who ultimately drove the British from the field. The small victories at Trenton and Princeton would prove decisive for American morale and radically alter the course of the Revolutionary War (both photographs, Library of Congress).

Nine. Revolution

British and another incredible, almost magical win for the Continentals. Victory and independence once again seemed now somehow achievable, and that spirit would continue to animate the Continental Army for five more long years of war. Yet it was freedom and independence for which the Continental Army marched, not booty, conquest, glory, or bloodlust; and in that regard it must be registered as a remarkable first in the long, bloody annals of human warfare. The methodology of war had not changed, but the reasons that men had taken up arms most certainly had, a fact that should give anyone pause.

In the insightful *War Is a Force That Gives Us Meaning* author Chris Hedges notes that all wars seem to have their causes, and that all causes are, in a sense, only empty fabrications, myths designed by those interested in pressing war to their own personal ends. "The cause is unassailable," he notes, "wrapped in the mystery reserved for the divine. Those who attempt to expose the fabrications and to unwrap the contradictions of the cause are left isolated and reviled."[17] While there is much truth to this, what Hedges does not seem to notice is that there are different sorts of causes, and that some are entirely justified. There is a great deal of difference, as an example, between the cause of emancipation that was embraced ultimately by many in the North during the American Civil War, and the cause of states' rights, which essentially meant slavery and was the cause on the lips of most Southern leaders of the period. Likewise, there is a tremendous difference between the cause, let's say, of exterminating human beings and the cause of preventing such exterminations. Many modern scholars and thinkers often appear incapable of distinguishing between these causes and are therefore incapable of explaining why some military efforts — say, to thwart the actions of a genocidal government — are morally proper, while others are not.

Generally speaking, the level of consciousness from which the cause emanates is the key. Ethnocentric, tribal causes emanate from level two, ethnocentric nationalistic causes from level three, but universal, compassionate, inclusive causes generally emanate from level four (and higher). In short, there is a clear difference between the causes generally embraced by each level; and liberty, democracy, civil rights, human rights, and the like are the causes championed only by level four and above. Those were, for the most part, the reasons for which the colonists fought the American Revolution. So while some causes, as Hedges points out, are little more than self-serving ethnocentric clamourings, others are not, and it is important to be able to distinguish between the two.

The reasons men marched during the American Revolution — their cause, if you will — were liberty, independence, and ultimately individual rights, and those reasons are on ample display in the documents of the period. The Declaration of Independence, as an example, penned in 1776 by Thomas Jefferson, is to this day considered one of the most primary and consequential documents in the history of democratic thought. As George Sabine notes, "The principles of the Revolutionary Era, first clearly stated by Locke and embodied in great political manifestoes like the American Declaration of Independence and the French and American bills of rights, summed up political ideals which in the nineteenth century seemed certain of progressive realization in the politics of all countries where the culture of Western Europe prevailed and might probably come to be realized throughout the world."[18] When Jefferson wrote that "we hold these Truths to be self-evident, that all Men are created equal, that they are endowed by their Creator with unalienable Rights, that among these are Life, Liberty, and the pursuit of Happiness,"[19] he was basing his assertions on the principle of natural rights; but natural rights had in themselves only recently become self-evident, and those to but a small handful of critical thinkers. After all, as previously pointed out, if these rights were in fact self-evident, why had they not been in greater evidence for the preceding thousands of years of humankind's civilized existence? This seems a fair question. And the answer is that, again, natural rights became self-evident only to the emerging level four mind. Thus before there was a significant emergence of level four there were no natural rights, no inkling that "Life, Liberty, and the pursuit of Happiness" should be considered self-evident aspects of human existence.

It is only at level four, it will be recalled, where the emerging mind becomes capable of taking the place of the other, thus understanding the world from someone else's perspective. Piaget termed this stage *formal operational*, and Ken Wilber provides insight into the conscious growth involved with revolutionary movements: "With the coming of formop [formal operational], the rules and norms of any given society can themselves be reflected upon and judged by more universal principles, principles that apply not just to this or that culture, or this or that tribe, but to the multiculturalism of universal perspectivism.... Socrates versus Athens. Martin Luther King, Jr., versus segregation. Gandhi versus cultural imperialism."[20] And, we might add, the colonists versus the British Empire.

This is the level from which the American Revolution sprang, and from which its rich list of revolutionary documents owe their genesis. The American Constitution and Bill of Rights, the writings of Thomas Paine — *Common Sense* and *The Crisis* — to name but two — are all creations of level four thought. The preamble of the Constitution reads, "We the people of the United States, in order to form a more perfect Union, establish Justice, insure domestic Tranquility, provide for the common defence, promote the general Welfare, and secure the Blessing of Liberty to ourselves and our Posterity, do ordain and establish this CONSTITUTION for the United States of America."[21] But this thinking can be found far beyond the great and thoughtful documents of the period. "When General Gage questioned the legitimacy of his rank," for instance, "Washington responded in a letter that was widely circulated in the American press: 'You affect, Sir, to despise all Rank not derived from the same Source with your own," Washington scoffed. "I cannot conceive any more honourable, than that which flows from that uncorrupted Choice of a brave and free People — the purest Source & original Fountain of all Power.'"[22]

The social and political ramifications of level four emergence were, during the American Revolution, still in their infancy, but over time they would become better defined.

> These ideals included the civil liberties — freedom of thought, of expression, and of association — the security of property, and the control of political institutions by an informed public opinion. Everywhere, as it seemed, these ends were to be practically realized by the adoption of the forms of constitutional government, by the acceptance of the rules that government must work within the limits set by law, that the center of political authority should fall within representative legislatures, and that all branches of government should be responsible to the electorate that tended to include the entire adult population.[23]

Although the American Revolution — and here I do not mean the war but the Revolution, of which the war was but a violent aspect — was at the time entirely unique to the North American continent, since the *potential* of level four consciousness lay dormant in every beating heart the possibility for the same sort of maturational leap remains alive across the globe in every population. As the psychiatrist Clifford Anderson notes, "Rather than resulting primarily from adaptation to the world, psychological development arises from a preexisting internal path.... If this is true, it means that within each newborn there is a program that, if correctly nurtured, will result in psychological maturity."[24] Given the proper conditions, then, the demand for freedom and some form of democratic government should naturally arise in every population once those proper conditions are in place.

But it must here be emphasized that democratic institutions arise naturally when populations are exposed to level four conditions, but either do not arise or wither at lesser levels. This is why democratic institutions have not traditionally dotted the globe, and why others regress once forced into existence in unsuitable situations. The relative ease with which Adolf Hitler dispatched the Weimar Republic government in pre–World War II Germany, for instance, serves as an example of this, which will be dealt with in more detail in upcoming chapters. But it is important to understand the simple fact that democracy cannot be force-fed to level two and three populations anymore than calculus can be force-fed to two year olds.

This subject obviously requires the utmost sensitivity, for it is open to misinterpretation. Indeed, at first it may appear to suggest that certain human populations are superior to others in that they are more "advanced," but this is not the case at all. In fact, the opposite is actually the case. The fact that *all* humans share the same potential confirms our most basic equality, and the fact that some populations are currently at different levels is a function of the maturational *conditions* in which those populations find themselves and has nothing at all to do with race, nationality, or ethnicity. Change the conditions and the levels will change — for better or worse. Thus it would seem that anyone interested in the advancement of democracy (and peace) on a worldwide basis should concentrate on the conditions that promote level four growth and not necessarily on the institutions of government themselves. This too will be discussed in greater detail in later chapters.

A print of Nathanael Greene originally printed for Seddon, Spotswood, Cist, and Trenchard, 1786. Greene replaced Horatio Gates as commander of the Continental Southern Army. Greene's Fabian strategy of generally avoiding a pitched battle with British general Charles Cornwallis would ultimately lead to the war's endgame at Yorktown, Virginia (Library of Congress).

The Revolutionary War ended in the fall of 1781 when Washington trapped General Charles Cornwallis' southern British army on Virginia's Yorktown peninsula. It was the ultimate result of a Fabian strategy (so named for the Roman general Fabius Maximus) employed by Nathanael Greene and his subordinate Daniel Morgan, two of Washington's most capable lieutenants. Greene, a Quaker from Rhode Island, had schooled himself in the art of war by reading everything he could get his hands on, while Morgan, whose background remains fundamentally unknown, emerged from the backwoods of Virginia's Shenandoah

Valley to become one of the most sage and gifted officers of the Revolution.

Greene, sent south to relieve Horatio Gates after that officer's debacle at the Battle of Camden in 1780, had adopted a fundamental strategy of hit-and-run, the Fabian strategy known at the time as a "War of Posts."[25] It was a realistic strategy considering that Greene's force was outnumbered by Cornwallis' professional army by nearly three to one. The idea was to attack then give ground, never standing up to a pitched battle, thus constantly harassing and frustrating the enemy, forever shifting the field of play. While the strategy — like most stratagems and battlefield tactics — had been embellished with a sort of intellectual veneer (a "War of Posts," if you will), in reality it was as old as the hills. Hunting packs, human and other, had dogged large prey while carefully staying out of goring range until the wounded animal collapsed from exhaustion since the hunt had existed, and the "War of Posts" was simply a modern, human rendition of that form of ancient hunt.

While Greene's strategy eventually worked wonders and was ultimately responsible for the war's endgame, it was his subordinate Daniel Morgan who displayed what might legitimately be termed military genius. Greene divided his smaller force and sent Morgan off to the west to try to confuse and harass Cornwallis. At Cowpens, South Carolina, Morgan baited and destroyed a British force sent to capture him led by the British wunderkind Colonel Banastre Tarleton, by means of a psychological and tactical deception so well conceived and executed that Tarleton's force — the cream of the Cornwallis light infantry — was wiped out almost to a man.

At the Battle of Camden American general Horatio Gates had placed his militia on the extreme left of his line of battle (directly opposite the most veteran British troops), and when they broke and ran, as militia had a natural tendency to do against advancing British regulars, his line dissolved, and the battle for the Continentals devolved into utter fiasco. Morgan keenly set his militia up in a slightly different format, one line out in front as skirmishers, another line of militia unseen on a down slope behind the first, and yet a third, main line of regular Continental troops behind that. Unseen, the cavalry was maintained as a general reserve behind the last line of infantry, ready to move where needed on a moment's notice.

Morgan simply asked his lines of militia to fire a few well-placed rounds at the closing British and then withdraw quickly behind the next line. When the charging British saw the American militia break and run they immediately had visions of Camden repeating itself, and they charged forward almost helter-skelter into the teeth of Morgan's trap. A misunderstood order on the Continental side almost turned Morgan's perfect plan into a disaster when one of his units mistakenly retreated, but he was quickly able to seize the moment and rectify the situation. The British regulars were gunned down in the trap, and Morgan quickly regrouped his militia and cavalry and led them on a double envelopment that eventually swept the field. The British military prodigy had been utterly destroyed by the American bumpkin in a defeat far more severe than Gates had suffered at Camden, and it was an effort that many would eventually term a military classic. Morgan's conception and execution at Cowpens stands as one of the few examples in the long history of warfare that might truly deserve the classification of masterpiece.

Morgan then rejoined Greene and the two led Cornwallis on a romp across North Carolina, until Greene turned back and the two armies collided headfirst at Guilford Court House. The fighting was severe, with both sides suffering heavy casualties, until Greene finally broke off the contest and withdrew. Cornwallis held the field, but his army had been so battered by the "victory" that he needed to both rest and refit his command. To do this he eventually removed his army to Wilmington, then on to a point along the Virginia coast where he hoped to be resupplied. That point was Virginia's Yorktown peninsula, and there he was ultimately cut off by Washington's forces on land and the French navy — now allied with the Continentals — by sea. With nowhere to turn, Cornwallis was forced to capitulate, and the Revolutionary War at long last came to an end.

The new nation would owe its military victory in large measure to George Washington, and over time much of its political makeup. While Washington was hardly an intellectual or, for that matter, even a writer of particular merit, his enduring faith in democratic government, the will of the people, and the nation he had helped form so informed his two terms as president that there is little question that the emergence of level four consciousness and the ultimate solidification of democratic institutions on the North American continent stand in his debt. Surely Washington, like all individuals, had his flaws, and these flaws have been well documented. He was, as an example, one of the wealthiest slaveholders in the colonies, but his contributions to democracy and the emergence of level four remain virtually unparalleled. He was breaking entirely new ground as president of a new republic, and his actions and footsteps set the standard for future generations in terms of, for instance, "the constitutional precedents he set for the executive branch in such specific areas as the cabinet system, control over foreign policy, the veto, executive appointments, and setting the legislative agenda."[26] In that sense George Washington became the archetypical president, and all others have since followed in his footsteps.

But there is little doubt that Washington's greatest postwar contribution came in the early spring of 1783, prior to the Treaty of Paris, when he headed off a potential coup by officers in the Continental Army stationed at Newburgh, New York. This incident has to a very great extent been lost to history;

Daniel Morgan (center right in buckskin) at General Burgoyne's surrender at Saratoga, New York. A backwoods brawler from Virginia's Shenandoah Valley, Morgan developed into a Continental officer of high skill and daring. At Cowpens, South Carolina, Morgan won a stunning victory against a crack force of British regulars by employing a brilliant tactical and psychological deployment (Library of Congress).

and while the true intent of the intriguers remains murky, it seems probable that some sort of an assault on the infant republican government was in the offing. The apparent plan was to march on the capital at Philadelphia and install Washington himself as the new head of the fledgling government; George Washington moved quickly and surely to head that effort off.

> Scholars who have studied the Newburgh Conspiracy agree that it probably originated in Philadelphia within a group of congressman, led by Robert Morris, who decided to use the threat of a military coup as a political weapon to gain passage of a revenue bill (the impost) and perhaps to expand the powers of the Confederation Congress over the states. Washington got wind of the mischief when he learned of petitions circulating among officers at Newburgh that contained veiled threats against the Congress if their pensions were not assured.... We can be sure that the crisis came to a head on March 11, when the dissident officers scheduled a meeting to coordinate their strategy. Washington countermanded the order for a meeting, saying that only he could issue such an order, then scheduled a session for all officers on March 16.[27]

Washington prepared a speech, then met with his officers at Newburgh in what was surely one of the most important and dramatic moments in the history of democratic government. A small part of what he said follows:

Nine. Revolution 109

George Washington: Commander of the Continental Army and first president of the United States. Washington's perseverance as a military commander and deep respect for democratic principles would guide the infant nation through its formation and early years (Library of Congress).

> And let me conjure you, in the name of our Common Country, as you value your own sacred honor, as you respect the rights of humanity, and as you regard the Military and National Character of America, to express Your utmost horror and detestation of the Man who wishes, under any specious pretences, to overturn the liberties of our Country, and who wickedly attempts to open the flood Gates of Civil discord, and deluge our rising Empire in Blood.[28]

It is said that his officers sat in virtual hushed reverence as he addressed them, some in tears as Washington fumbled with his spectacles. Many finer speeches have been given in the name of democracy, but none more heartfelt, and certainly none more critical. With the deliverance of Washington's speech the coup simply dissolved into thin air; no one would lift a finger against his wishes or his defense of republican government. "In this culminating moment of his military career," as Joseph Ellis puts it, "Washington demonstrated that he was as immune to the seductions of dictatorial power as he was to smallpox. And, as was so often the case with his most dramatic decisions, the reasons for his behavior were so deeply buried in his character that they functioned like a biological condition requiring no further explanation."[29]

Under somewhat similar circumstances in 1782 King George III, upon hearing that Washington might well pass on being crowned "king" of the new North American nation, exclaimed that, if this were true, George Washington would be "the greatest man in the world."[30] Not only would Washington pass on power once, he would walk away from power repeatedly, leaving to the nation he had fought to conceive a legacy of executive restraint and respect for constitutional law historically unmatched. "Whereas Cromwell and later Napoleon made themselves synonymous with the revolution in order to justify the assumption of dictatorial power, Washington made himself synonymous with the American Revolution in order to declare that it was incompatible with dictatorial power."[31] That stance laid the foundation for the peaceful, civil transfer of power in the United States that is to this day the envy of the world, yet is that something taken completely for granted by most citizens of the modern republic? It is not a coincidence that the capital of the new republic was named Washington.

The American Revolution was without question one of the signal events in the affairs of *Homo sapiens sapiens*, the spontaneous leap into a whole new level of culture, thought, and government. But it was just a beginning, and the smallest beginning at that. While the motives and documents that fed the revolution were true level four, much of the infant nation remained at level three, and level three causes and aspirations would continue to motivate and bedevil Americans for quite some time to come. But growth is always halting and spasmodic, thus many of the great causes of the nineteent century and even the early twentieth century — Manifest Destiny and the predatory Mexican and Spanish American Wars, as examples — would be products of this latent level three ethnocentric outlook. The American South, wedded to the institution of human bondage as a means of production, would over the years remain firmly in the level three cultural camp. In time, the rising tide of level four in the industrial North and this southern, regressive level three outlook would clash in a tempest on the North American continent known as the American Civil War.

Ten

A Violent Clash of Levels

So far this inquiry has covered considerable ground, and along the way quite a few pieces of the puzzle concerning the nature and origin of predatory warfare have fallen into place. It may be wise at this point to take a step back and look them over again.

We have discovered that war erupted sometime around 10,000 B.C. as a result of the sudden, far more comprehensive understanding of death by the emerging level three mind. We observed how that comprehension was in turn violently projected, and that this process made war seem somehow natural, indeed, even necessary to those early pioneers of the level three landscape. It has also been demonstrated how the developmental model for level three thus became the warrior king, and that this psychological investment created the original warrior archetype, which led to widespread warfare throughout humankind's next few thousand years on earth.

We noted that the birthplace of level four lay with kings and their retainers, and that as a result the emerging level four ego was in a very real sense born to war. We observed how that original ego, knowing no natural boundaries, often set out to conquer the world, and we saw how Alexander — surely one of the greatest egos the world has ever seen — impressed the Homeric codes upon the warrior archetype, the archetype that remains with us still. Thus did these original human behaviors become patterns; and as more and more individuals emulated these behaviors, the patterns became ruts; and as the centuries passed *Homo sapiens sapiens* seemed somehow stuck in a violent world that appeared to be beyond their ability to either fathom or control.

Recall as well that as level four spread and matured it was able to take the place of the other for the very first time, and from that fundamental comprehension in time would spring human rights, democracy, and a natural tendency toward peace — the first real break from the ruts of war. We also observed how the level four mind would create the scientific method, usher in the Renaissance, and dazzle the world with the Enlightenment. The new science of experimentation would in time produce an industrial revolution that would soon turn its attention to the implements of war, creating an arms race that would in only a few centuries spiral completely out of control.

It has been noted that level three tends to be rigid and imperial, agrarian in its culture, religiously fundamental, and socially stagnant, while level four, on the other hand, tends to be passive, industrial, religiously tolerant, and socially upwardly mobile. Roughly speaking, if we think of rural China as level three and suburban Great Britain as level four, the picture will start to come into focus. Recall also that these two developmental levels, being drastically different, have historically never understood one another.

Finally, we observed how the emerging collective level four ego expressed itself through the development of the modern nation-state — a process that in many locales around the globe is still ongoing. This brought this inquiry forward from the year 10,000 B.C. to roughly the nineteenth century, and to a time when all of these forces seemed to coalesce like converging tornadoes in an explosion on the North American continent called the American Civil War. So to grasp how these factors interact, and to better understand the current situation across the globe, our inquiry will take a close look at the American Civil War from the deeper perspective of human development.

The Civil War was without question the most traumatic experience the United States has ever endured. It is estimated that over 620,000 men lost their lives in the war — more than all other American wars *combined*, with perhaps two-thirds of those deaths due to illness and disease.[1] No one knows for sure just how many men were wounded, but in most Civil War engagements the wounded tended to total some three to four times the number of battlefield deaths.[2] So it would be safe to say that perhaps as many as another 900,000 men were wounded in battle and returned home in some way scarred for their efforts. By war's end over 3 million men would serve in the land armies representing both sides — almost 10 percent of the nation's total prewar population.[3] The war demanded a staggering sacrifice in both human and financial terms. The North went deeply into debt as a result of the war, while the South's economy was destroyed along with much of its infrastructure.

The American Civil War began in April 1861 when Confederate forces fired on Fort Sumter in Charleston Harbor, and few at the time foresaw the extent of the horror that was fast approaching. In fact, both sides initially imagined that the war would be over in a matter of months only and the troops safely home by Christmas. Indicative of this attitude was Abraham Lincoln's initial call for 75,000 volunteers to serve for a 90-day period. It was imagined in the White House that the war would be over in only three months.

After the Union's first disastrous defeat at Bull Run in July 1861, however, Lincoln began to realize the magnitude of the effort that would be required to subdue the South, and he called for an additional million men to serve for a full three years. When Confederate forces first fired upon Fort Sumter, the Federal government's entire land force consisted of only 16,376 men and officers.[4] Over the next four years some 2.2 million men would serve the Federal cause, an increase in size, scope, and effort for the still comparatively young nation that was nothing short of breathtaking.[5]

The cause(s) of the war have been long debated, and, arguably, few would disagree today that slavery was the core issue. Confusing this conclusion from time to time is the fact that many Northern soldiers insisted they were fighting not for the abolition of slavery but only to preserve the Union, and additionally that the North's industrial plant stood to gain handsomely from any prolonged encounter. Thus the war has occasionally been interpreted as essentially a profit-taking ploy instigated by greedy Northern industrialists, or else an attempt to do nothing more than preserve the political status quo. But all three of these interpretations fail to fully appreciate the full range of motivations inherent in the conflict, for all three are in fact core elements of level four emergence.

As pointed out above, level four tends to be passive, industrial, religiously tolerant, and socially upwardly mobile. Since level four is the first level that can objectively take the place of the other, it also represents the first worldview that would extol the virtues of individual freedom and democratic government. As Ken Wilber explains, this was a first for humankind: "The values that liberal Westerners tend to share, the values of the Enlightenment (the values of rational industrialism)—namely, liberty, equality, and freedom— were simply *not* the values of *any* other societal type, *ever*."[6] In that sense, level four thought contained the first genuine threat to the institution of slavery, but this level also *included* a strong sense of nationalism, *along with*, and based upon, an industrial economy. The three were not separate, independent phenomena, but rather objective evidence of the same level four emergence.

The American Civil War, fundamentally understood, was a clash between an emerging level four North, and a stagnant, agrarian, level three South, a region with precious little industry and wedded to the institution of slavery as a means of production. Southerners wanted to do nothing more than preserve their way of life, and thus they argued that the war for them was one in defense of "states' rights." This is why the war's cause cannot be reduced to slavery alone, for it is true that a welter of other rights and issues also consumed Confederate politicians. But, of course, as Abraham Lincoln correctly pointed out, the right and freedom the South wanted particularly to defend was the right and freedom to enslave other people. So in many ways the two regions were fast becoming opposites, as might be expected, but nowhere was the prewar gulf between North and South greater than in the field of education.

As has been discussed, historically education had been the exclusive province of the very rich and powerful. Alexander may have had Aristotle as his personal tutor, but most Greeks barely got by. As we saw in chapter eight, at the time of Martin Luther's trial (A.D. 1521) only 5 percent of the German population could read and write. But all that was about to change with the American experiment in democratic rule. "Before the Civil War, the North was the center of education in the country. By 1850, New England was a world leader in educational facilities. Ninety-five percent of New England adults could read and write, and 75 percent of the children were in school."[7] From a historic perspective these are simply astonishing statistics, but that was not the most important innovation. Years before the Civil War began, "every Northern state had some form of mandated public school system, but the South did little to support its own educational and cultural development."[8] Public education, the open door to level four development, had arrived in earnest in the North, but the South remained a far different story.

The Civil War has been analyzed from many different angles, but rarely has anyone taken a hard look at these disparate, prewar educational systems and drawn some basic, developmental conclusions from the data. "In 1860, 66.8 percent of white children and 34.9 percent of black children in the North were in school, compared with 43.9 percent of white children and 3.2 percent of free black children in the South."[9] In several Northern states more than 50 percent of black children attended public schools. "Thus, in much of the antebellum North, an African American child had a greater chance of getting an education than a white child did in most of the antebellum South."[10]

Thus public education, an emphasis on literacy, and large, urban settings greatly facilitated the emergence of level four consciousness throughout the North, while the exact opposite was taking place in the South. "New England's emphasis on education created a regional culture of innovation, enterprise, and experimentation."[11] Not surprisingly, it was New England that led the way in the push toward the emancipation of slaves and the ultimate abolition of slavery.

In the years immediately preceding the Civil War every city in the North had one daily newspaper or more. Those papers exposed their readership to all sorts of new and radical ideas, from across the country and across the world. In the South, on the other hand, newspapers were literally feared for the ideas they espoused. "By 1860, the South had become a closed society when it came to discussions of race, slavery, and public policy. Southern states banned the circulation of books, newspapers, or pamphlets that criticized slavery or questioned its morality."[12] Throughout the South, simply carrying a book such as Harriet Beecher Stowe's *Uncle Tom's Cabin* became a criminal offense. The issue of the day across the United States was the "curious institution" of slavery, but in the South there would be no public discussion at all.

That curious institution, of course, was by no means a uniquely American development, although the Western hemisphere would certainly stamp it with its own, racial thumbprint. As Ken Wilber explains, "*All* types of pre-industrial societies had slavery, with no exceptions — foraging (hunting and gathering), herding, horticultural, maritime, and agrarian."[13] Some Native American tribes routinely took and held slaves long before the arrival of Europeans. The Iroquois, for instance, maintained three classes of slaves: "The first were admitted into families, and sometimes became chiefs, though still considered slaves. The second were given to richer Indians, and had food and shelter, but nothing more. The third were young women and girls, continually exposed to every danger. Often, however, they were saved from death to become wives. As slaves the treatment of these girls depended on the temper of her mistress, and this was often cruel."[14]

The institution of slavery in Europe, the Middle East, and Africa existed centuries before the discovery of the New World.

> Certainly, servile labor in Africa was an institution that had long preceded the appearance of the white man. Captives taken in war invariably became slaves. Others were kidnapped or else enslaved as a punishment for crimes.
>
> Nor were Europeans the first outsiders to come in quest of African slaves. Arab merchants had been transporting their own human purchases across the Sahara for some 700 years before the European traders arrived in Guinea; over the course of the centuries, more than three million people were exported to Islamic states on the Mediterranean and Red seas, and innumerable others perished on the scorching march through the desert.[15]

Moral qualms regarding slavery would not arise until level four consciousness began to emerge, most notably around the seventeenth and eighteenth centuries, and in the North American colonies those budding moral qualms would at first be feeble and inconsequential.

In the American colonies the first real objection to slavery came from the Quakers in the late seventeenth century, but those objections would grow far louder over the following years. Congress, in accordance with the Constitution, outlawed the African slave trade in 1807, and by the midpoint of the nineteenth century many groups, associations, and societies openly campaigned for abolition. Yet it was not so much slavery as it was the extension of slavery into the Western territories that brought the issue to the boiling point in Kansas and Missouri. Those bloody conflicts would radicalize the abolition movement even more and bring Abraham Lincoln to the White House on an abolitionist platform in 1861. Lincoln's election was the smallest spark that finally lit the fuse to a vast, simmering powder keg.

In purely psychological terms, the appearance of inclusive public education along with the increase of regional media in the North forced the South into what might be classified as a state of denial regarding the topic of slavery. Any view in opposition to its level three devotion to the curious institution was literally repressed. Meanwhile, conditions in the North were becoming well suited for the rapid development of level four consciousness, while the South remained not only backward in terms of its educational structure but also backward in terms of its worldview, attempting to censor and criminalize its way out of a reality that was approaching with a fury. These two components were the subterranean fuel, so to speak, that would drive the Civil War far beyond anyone's early expectations in terms of ferocity or duration. The war was literally a clash of past (level three) and future (level four) civilizations, a fact noted with remarkable insight by Confederate general Edward Porter Alexander long after the war had ended when he wrote that the Confederacy had been defeated by far more than Federal arms: "but the trouble was that we were struggling against changes which the advance of the world in railroads & steam-boats & telegraphs, in science & knowledge and commerce, &, in short, in civilization, had rendered inevitable."[16]

Adding significantly to the war's violence — to the butchers bill, as Lincoln would later put it — was the initiation of the technological arms race, an arms race that is, to a great extent, with us still. Of course, weapons had always been improved and tinkered with over the course of history, but with the advent of the level four industrial revolution the rate of weapons development would increase exponentially. Because of the increasing lethality of small arms ordnance, the Civil War would, in four agonizing years, rapidly and painfully evolve through an entire century of tactical development. Both armies would begin the war at Bull Run fighting in the old Napoleonic style, marching large infantry concentrations across open fields of fire in order to deliver a concentrated volley. Splendidly Homeric, this tactic was reminiscent of the European battles of Austerlitz or Waterloo in the earlier part of the century. The war then ended in the trenches surrounding Petersburg, a tactical situation eerily reminiscent of the engagements at Verdun, Belleau Wood, or the Somme during World War I.

As Robert O'Connell points out, virtually all of the major officers on both sides of the Civil War — Lee, Grant, Jackson, Longstreet, McClellan, and so on — received their early military training during America's war with Mexico in 1847.[17] At the time, the standard-issue firearm was a smoothbore musket with a range of perhaps 400 yards, and of limited accuracy.[18] Against such weapons the Napoleonic line of battle worked well and with comparatively meager casualties.

By 1861, however, level four science had already spawned the industrial revolution. In 1793 Eli Whitney's cotton gin introduced laborsaving machinery to North America, and from that moment on the technological race switched into high gear. The cotton gin would greatly impact cotton profits across the South, further isolating the southern, level three mind from any thought of abandoning slavery, while in the North it spawned large-scale industrialization. That industrialization would soon turn its attention to issues of military ordnance.

By the beginning of the American Civil War the standard-issue firearm for the Federal infantryman had changed from the smoothbore musket of the Mexican War era to the Springfield 1861 rifle.[19] The Springfield fired a conical bullet developed by the French army officer Captain Claude Minié.[20] Later dubbed the minié ball in the United States, the soft lead ball worked well in a rifled barrel, dramatically increasing both the range and accuracy of the weapon. The 1861 Springfield could fire a .58 caliber bullet 1,000 yards with reasonable accuracy, and the result on the field of battle was simply horrific. Virtually overnight the old Napoleonic tactics were rendered obsolete, but for the first three years of the Civil War no one in authority seemed to take notice. Troops were still marched headlong across open fields of fire into the teeth of enemy positions. The result on the field of battle, notes Robert O'Connell, was mass slaughter.[21]

Confederate infantry assault, known as "Pickett's Charge," on the third day of combat at Gettysburg. Improvements in the standard-issue infantry rifle and the coordination of artillery fire would make such large-scale assaults increasingly ineffective (Library of Congress).

Ten. A Violent Clash of Levels

Portion of Confederate trench system at Petersburg, Virginia. The American Civil War would pass in four years through a virtual century of tactical evolution, from the Napoleonic line of battle to the trench systems characteristic of World War I (Library of Congress).

Absolutely no one, military or civilian, was prepared for the sort of carnage that arose from this unexpected imbalance of firepower and tactics, and it would take almost four years for people to react. Military medicine, a primitive institution to begin with, was now simply overwhelmed. Thousands of men were left to die on the field of battle, and in some cases (Second Bull Run and the Wilderness as examples) many of the dead were never buried at all. Individual infantrymen, painfully aware of the suicidal nature of the combat they were being called upon to endure, began digging rifle pits and piling rocks or felling trees for protection at the first opportunity; but the war would not truly descend into the trenches until the battle of Cold Harbor in 1864, and later the siege of Petersburg.

Yet, despite improvements in the coordination and accuracy of artillery fire and the extraordinary increase in the lethality of small arms, during the war's first three years no volume of fire or number of casualties seemed to make an impression on the general officers in charge. It was simply business as usual, no matter how many men died in the process. The only sensible tactical alteration was the relegation of the cavalry — once the thundering death blow on the field of battle — to the auxiliary arena of reconnaissance and flank protection, a demotion from which it would never recover.

Exemplifying this inexplicable myopia was the death of Federal general John Sedgwick. Sedgwick, a graduate of West Point and a veteran officer with 27 years' experience under his belt, was the ranking officer of the Sixth Corps, and one of the most respected generals in the Federal army. In May 1864 at Spotsylvania Court House while repositioning artillery, Sedgwick came upon some of his men hiding behind trees and lying low in gullies due to enemy fire. The general, in true Homeric style, began to gently chide his troops for what he considered an irrational fear of the enemy. Why, the Rebels, he explained while standing out in plain view, "couldn't hit an elephant at this distance."[22] After almost three continuous years of war it would seem that Sedgwick should have known better, but the obvious capabilities of the modern rifle and the many sharpshooters who carried them were apparently yet to dawn upon him. He had no more than finished his last sentence when a bullet passed through his brain, and he fell dead on top of one of the men he had been teasing. As his body was carried off by his stunned staff, the troops went quietly back to their cover.

Civil War technology would produce repeating rifles like the Henry, capable of firing sixteen bullets without being reloaded.[23] Those weapons would represent a remarkable leap in technology, and for the field of battle they were an ominous leap in firepower. The slaughter — already horrific — was now accelerated by a potential factor of 16. Fortunately for the Confederate soldier on line, so enamored was the Federal military establishment with the single-shot Springfield that the majority of Henries and other repeating carbines would be distributed only to the cavalry, and that enormous advantage in firepower would be squandered.

Numerous modern inventions revolutionized the art of war during the four years between 1861 and 1865. The use of trains to transport men and supplies, the telegraph, balloons for intelligence gathering, the first primitive submarine, the Gatling gun (machine gun), and conscription would all become national fixtures during the next century, and all were evidence of the level four mind hard at work. On the water the ironclad was introduced, and in that regard no one event would have more significance than the clash at Hampton Roads between the *Monitor* and the *Merrimac*. For as Robert O'Connell notes of the coming arms race, the balance of power would suddenly be married to the ever-grinding process of technological innovation and the frightening potential that a single new weapon might overnight render an opponent's entire arsenal obsolete.[24] For all practical purposes, that is very close to what happened on the waters off Newport News, Virginia, an event that initially touched off near pandemonium in the Lincoln White House, and ultimately produced the modern arms race.

The great potential (and simultaneously the great fear) of military technology is that one side will develop a weapon or system that simply trumps all others, rendering the other side's entire arsenal instantly obsolete. For years that has been the holy grail, so to speak, of military technology, and that goal the natural driving force behind the arms race. But the fundamental problem with this race, as O'Connell makes clear, is that any momentary advantage gained seems to be rapidly checked with another equal, or even superior, innovation. Thus the race goes on and on, neither side ever gaining the long-standing advantage they initially sought.

The first significant instance of this dynamic at play was the drama surrounding the clash of the ironclads *Monitor* and *Merrimac* in 1862. While many people are familiar with the slugfest between the two boats on the waters of the Chesapeake Bay, few are familiar with the day that preceded it. Yet it was one of the most momentous in American military history, the reason being that for a very short period of time it appeared the Confederacy had actually achieved that most longed-for of military fantasies — the invincible weapon. Here's how it went.

The Rebel ironclad had been the brainchild of Stephen Mallory, the Confederacy's secretary of the navy. Grasping the enormous superiority of Northern industrial might, Mallory reasoned he could never match the Federal navy boat for boat. Prior to the war, however, France had already developed a few ironclad floating batteries, and it was from their example that Mallory drew his inspiration. His hope was to create a new ship design that single-handedly could raise the Federal blockade and carry the war to the North.[25]

When Virginia seceded from the Union in 1861 the Federal navy had abandoned one of its two deepwater shipbuilding facilities, the navy base at Norfolk on the Chesapeake Bay. Much of the base had been destroyed and the ships under construction set ablaze. The *Merrimac*, a Federal frigate under construction at the time, was located by the infant Confederate navy, raised, stripped, and converted to the new ironclad design. When the work was finally completed in March 1862 she was rechristened the *Virginia*, but few, including her own crew, would ever know her as anything other than *Merrimac*.

On March 8 the ironclad slipped from her mooring in Norfolk and headed out to contest the waters of the bay with the wooden warships of the blockading Federal fleet. The contest proved to be a complete mismatch and was one of the worst days in United States naval history. The *Merrimac*, boasting four specially designed Brooke rifles fore and aft and three Dahlgren smoothbores on each broadside, was a formidable weapon, despite several significant design flaws.[26]

It was early afternoon when the *Merrimac* came upon the Federal frigates *Congress* and *Cumberland* anchored off Newport News. She steamed into direct broadside range without firing a single shot. The Federal blockaders, stunned by the audacity of the *Merrimac*'s approach, responded immediately. "The *Congress* then fired a full broadside point-blank. Nine hundred pounds of hurtling metal smashed fully home. In horror the gunners saw the balls ricochet into the air like pebbles off a roof."[27]

The *Merrimac* came about and replied with a broadside of its own, tearing the *Congress* to pieces. Then the ironclad edged around and headed toward the *Cumberland* lying in wait nearby, with every gun run out and loaded. But the *Merrimac* approached head on, avoiding the *Cumberland*'s broadside, and rammed the frigate at the bow. Then both ships exchanged furious broadsides for the better part of an hour as the *Cumberland* slowly sank into the waters of the bay. As the *Merrimac* turned to finish off what remained of the *Congress*, Federal tugs began pulling the last three giant ships of the Federal fleet—*St. Lawrence*, *Minnesota*, and *Roanoke*—south toward the action, only to run them aground in all the excitement.

While the skipper of the *Merrimac* would liked to have closed with the grounded *Minnesota* and methodically finished off the remnants of the Federal fleet, daylight was now fading, and the ironclad had been somewhat damaged in the fray itself. So it was decided to make for the Norfolk yard, then return the following morning to destroy the *Minnesota* and the other stranded Federal ships. For all intents and purposes, by day's end the Federal fleet was a shambles.[28]

The following morning, Sunday, March 9, 1862, a scene occurred in the White House the likes of which has rarely been witnessed, either before or after. Graphic reports had been received earlier that morning recounting the destruction of the fleet and attesting to the *Merrimac*'s seeming invulnerability. President Lincoln responded by calling an emergency meeting of his cabinet. The president, according to Gideon Welles, then secretary of the navy, was so upset he could hardly speak. Other cabinet members simply sat by, entirely dumbfounded. The scene could be described only as one of utter panic. According to Welles, it was "the worst moment of the war."[29]

The secretary of war, Edwin Stanton, was reportedly frantic. He ran through the White House, flinging open windows, half-expecting to see the *Merrimac* steaming up the blue waters of the Potomac River to begin a bombardment of Washington City. What, after all, was to stop it? "The *Merrimac*,

Ten. A Violent Clash of Levels

he railed, would sink every vessel in the navy, capture Fort Monore, cut off Burnside in the Carolina sounds, retake Port Royal, and lay New York and Boston under 'contribution.'"[30] The Rebels had, from the perspective of almost everyone in attendance, developed a weapon for which the Federal arsenal had no answer. If this were true, the war appeared lost.

Unknown to everyone present except Welles, however, the small Federal ironclad *Monitor* — rushed to completion due to reports of the *Merrimac*'s looming availability — had arrived off Newport News just as the *Merrimac* had turned to steam back to Norfolk the preceding day, and there it took up a position in defense of the stranded *Minnesota*.

That Sunday afternoon the two ironclads would blast away at one another for hours in true Homeric fashion, neither side gaining an advantage, until the *Merrimac*, running short of both fuel and ordnance, broke off the contest and returned to base in Norfolk. Stephen Mallory's fond hope of developing a weapon that might trump the entire Federal arsenal and win a breakthrough war for the Confederacy had become a stunning reality — but only for a single afternoon. Technology had responded, and it would be the gritty little *Monitor* with its swiveling turret that would actually prove the model for naval designs of the future.

Over the coming decades many others would try to reproduce a more durable version of what Stephen Mallory had managed to achieve for only one day. Dreadnaughts, machine guns, tanks, airplanes, and poison gas, among many other weapons and weapons systems, would all emerge from the technological arms race, ultimately delivering *Homo sapiens sapiens* to the very brink of nuclear annihilation. In less than one century (the blink of an eye when compared to the 12,000-year history of war) science and technology would take humankind from the 1861 Springfield with its one minié ball and killing range of 1,000 yards, to the intercontinental ballistic missile capable of utterly destroying vast regions from halfway around the globe. Humankind had gone naively in search of the invincible weapon, but stumbled instead upon a seemingly ungovernable formula for apocalypse.

Fortunately, a few sober souls would notice that the arms race, almost blindly conceived to improve the implements of war, had also delivered humanity to the very doorstep of war's practical and intellectual obsolescence, because these new nuclear weapon systems no longer offered a tactical edge but only the edge of a cliff with no bottom in sight. The use of these weapons no longer augured victory, or triumph, or even mere heroism, but obliteration.

The American Civil War has been called the first truly modern war. By this, it is generally meant that the war became total, the objective being the farms that produced the food that fed the enemy as much as the rails that carried the food or the enemy's soldiers who ultimately ate it. The line between civilian and combatant became blurred just as the old Homeric codes, now increasingly meaningless due to the sheer lethality of the battlefield, were becoming increasingly obsolete.

In that sense the American Civil War teaches a great deal. Like a crystal ball, the face of what war would look like in the coming twentieth century was contained in the War Between the States for all to see; but if history is any guide, very, very few absorbed the lesson. Level four technology would increasingly come to dominate the modern battlefield, while Homeric qualities such as pluck, fortitude, and heroism would be rendered increasingly meaningless. Suddenly a thousand brave souls could be gunned down by a few hidden machine guns, and in such a world heroism would be reduced to a worthless commodity. Increasingly, the bravest were the first to die, often having accomplished almost nothing of value by means of their deaths. An industrialized world would simply turn war into an enormous meat grinder where men would be ground up by the millions, and the mere idea of heroism would be turned into a sad sort of farce.

So the horrors would be repeated, and not symmetrically the next time around but exponentially. At Gettysburg, for instance, the American Civil War's most costly battle, some fifty-three thousand men would go down as casualties in three terrible days of combat. Yet only 53 years later the battle of Verdun would drag on for a horrific 302 days and generate almost 1,000,000 casualties. And sadly, that was just the beginning.

But the American Civil War would represent far more than merely an advance in weapons technology and tactical innovation. The war was in the end much more about the expansion of level four ideas than it was level four battlefield ordnance or the ascendance of trench warfare, and the documents of the period clearly establish this oft-forgotten fact. The Civil War changed not only the fundamental civil rights of the nation's population but the very course of the nation's future. In that sense the writings generated during the Civil War reveal a great deal about the continued emergence of level four thought into Western culture, and they therefore offer firm testimony as to the evolution of war.

Eleven

Documents of Change

The Civil War continued the emergence of the radical, level four ideas that had burst upon the scene some eighty years previously. The American Revolution had ushered in not only a new form of government but also a new way of conceiving of government, a sea change in the social, cultural, and political landscape of *Homo sapiens sapiens*. That revolutionary event laid the groundwork for what would eventually become the tenets of modern liberalism: democratic rule, individual freedom, tolerance, liberty for all, among others. At the same time, however, the American Revolution, sweeping though it had been, was in reality just a beginning, the initial significant burst of level four thinking, and that initial burst naturally arrived in limited form. While Thomas Jefferson proclaimed in rather windy tones, for instance, that all men had been created equal, the truth of the matter was that in the United States many men, women, and children were not equal at all. Rather, they had been born or sold into bondage, and the practical and moral dilemma posed by the institution of slavery had simmered just below the American social and political surface for decades.

For southern politicians the debate over slavery had been simply too difficult to deal with, thus had the topic been largely swept under the rug since the original Constitutional Convention. Rather naturally, however, with the tide of level four consciousness rising steadily in the American population, the topic refused to go away. In 1790, for instance, "two Quaker petitions, one arguing for the immediate end of the slave trade, the other advocating the gradual abolition of slavery itself, provoked a bitter debate in the House. The petitions would almost surely have been consigned to legislative oblivion except for the signature of Benjamin Franklin on the second, which transformed a beyond-the-pale protest into an unavoidable challenge to debate the moral compatibility of slavery with America's avowed revolutionary principles."[1] Washington, while agreeing in principle with Franklin regarding slavery, refused to lend his considerable support to the petition, and there the matter rested. "He [Washington] endorsed Madison's deft management of the debate and his behind-the-scenes maneuvering in the House, which voted to prohibit any further consideration of ending the slave trade until 1808, as the Constitution specified; more significantly, Madison managed to take slavery off the national agenda by making any legislation seeking to end it a state rather than federal prerogative."[2] Thus had back room maneuvering put an end to the debate over abolition and the moral question of slavery itself. But under-the-table shenanigans could hardly put an end to the problem, which would fester for decades until finally exploding into violence in the Western territories of Kansas and Missouri.

In that sense it can be seen that the infant republic, while certainly revolutionary in many constructive ways, simply had not lived up to the rhetoric of its own principles nor the documents through which those principles had been expressed. The Declaration of Independence might have declared that all men were created equal, but for many the reality of life in the United States was something radically different. This glaring omission — the unwillingness to address the issue of slavery — was in fact the 10,000-pound gorilla in the room for the first eighty years of the republic's existence, and a gorilla that would break a great deal of furniture before it was finally, and belatedly, addressed. As developed in the preceding chapter, the South would over the years slip into deeper and deeper denial over the situation, refusing to confront this profound contradiction in its midst, while the North, fundamentally absent slavery, continued to move deeper into level four thinking. Some form of clash appeared inevitable.

The American Civil War, then, is best seen as a spontaneous clash between a growing, level four North and a static, level three South, a region wedded to slavery as a means of production, and slavery an institution that was already culturally, morally, and economically indefensible. While war outright actually came as a shock to many — both North and South — and thus the initial war aims of the Federal government simply the restoration of the republic to its pre-conflict status, it would not be long before the deeper, far more profound goal of abolition would be embraced as a national objective. That this goal was not initially supported by all should not come as a surprise.

While the initial revolutionary principles of the United States clearly had their origin in level four patterns of thought, in 1861 the vast majority of the population of the United States both North and South remained firmly arrested at level three or lower.[3] Dr. Clifford Anderson makes this clear: "Historians with an eye for psychological data generally agree that before this century [the twentieth century] the most common life cycle profile was childhood followed immediately by adulthood. Although most people presumably achieved physical maturity, they probably did not mature psychologically beyond childhood [level 2–3] as a stage."[4] The upshot of this was a population firmly rooted in concrete or literal patterns of thought, generally incapable of abstract or relativistic thinking and therefore unable to apply the broad principles of freedom, democracy, and civil rights to people beyond their own narrow groups and borders. While many in the North, therefore, could grasp the issue that the Union needed to be "restored" and as a result could readily formulate a conception of secessionists as "traitors," far fewer could apply the principles of freedom to slaves in the far-off South whom they had never met or seen. It would therefore require a virtually Herculean effort to shift the war aims of the North from simply restoring the Union to emancipation, and this task fell fundamentally upon the shoulders of Abraham Lincoln, then president of the United States.

The Revolutionary War had been fought to establish a new nation predicated upon the emerging level four dictates of universal freedoms, but the final result had been lacking because the issue of slavery (among others) had not been sensibly addressed but rather swept under the carpet in what can only be described as a willful example of cultural and political denial. During the course of the Constitutional Convention the topic of slavery had been simply too hot to handle, and northern politicians acquiesced to southern intransigence, fearful the issue would sabotage all efforts at reform, leaving the infant nation hopelessly divided and subject to failure as a result. This denial, of course, accomplished nothing constructive. The issue finally erupted into violence in the Western territories as pro-and antislavery forces clashed bitterly over the extension of slavery into those regions, both forces furiously mindful of the congressional voting imbalance that might have been created with a radical swing in either one direction or the other. What for eighty years could not be openly discussed had finally exploded into bloody conflict.

The growing movement for abolition in the North that had preceded the Civil War was a belated effort by the growing level four mind to correct the glaring omission that had flawed the original republican government. The effort, due to the developmental level of most of the population, would require a determined and sweeping process of change to succeed. That process is clearly to be seen in the documents of the time, documents that, in a series of logical steps, shifted popular opinion from apathy or outright resistance to acceptance of abolition by means of a constitutional amendment. When considering the cultural and psychological resistance involved, this represents no small achievement.

The Emancipation Proclamation is the first document to be examined in terms of its level four impact, although Abraham Lincoln had a long history of arguing against slavery before he became president and therefore a long paper trail of antislavery material to choose from. The Emancipation Proclamation remains one of the most important documents of the period, and probably one of the least understood. To understand the document it is essential to understand just how Lincoln viewed the conflict between North and South and his subsequent role as a result of that view.

Lincoln never accepted the legitimacy of the Southern Confederacy. To him the secessionist government was simply an insurgency with no legal standing at all; thus the conflict was not a "war" in any legal sense, but an illegal takeover of part of the country by illegitimate intriguers. This notion was based on the fact that the United States had in fact preexisted in some manner the Revolutionary War and had been conceived as a union by the people; it could therefore be broken apart only by the consent of the people as a whole, and not simply by the various state legislatures of a given region.[5] From this perspective Lincoln viewed the legislative act of secession in the various southern states as illegal, and the Confederacy, therefore, as an instrument of insurgency and not the will of the people. Lincoln thus still considered himself president of all the states — both North and South, whether seceded or not.

Lincoln was a lawyer, and his was most certainly a legal argument. While the question of secession will presumably never be adjudicated in any modern court of law, thus a definitive ruling on the issue never achieved, Lincoln's position was nevertheless anchored in fact and was therefore at least a valid legal hypothesis. Working from this proposition, then, Lincoln beheld his duty as president to be the suppression of the rebellion — this illegal insurrection — very much as Washington had put down the Whiskey Rebellion during his own administration years earlier.

Being a lawyer, Lincoln also understood the legal constraints imposed upon him as president by both the Constitution and prior case law. When Lincoln took office in 1861, slavery was very much a protected constitutional right under the laws of the United States. Indeed, in the case *Dred Scott v. Sandford*, decided by the U.S. Supreme Court in 1857, the court determined that a slave had no rights whatsoever under the Constitution, and further that slaves had no legal standing beyond that of physical property.[6] Lincoln himself had never argued that slavery was illegal, only that it was immoral, and when he took the oath of office to uphold the Constitution he pledged as a matter of course to defend the institution of slavery — at least in the Deep South where it remained the law of the land — whether he liked it or not.

In that sense, Abraham Lincoln had no power whatsoever to emancipate anyone anywhere in the United States from bondage under the executive powers granted him as president of the United States, and this was a legal constraint he perfectly well understood. This fact seems, at times, lost on modern observers of the Civil War, thus the Emancipation Proclamation is often portrayed in modern analysis as little more than a half-hearted maneuver, executive blunder, or crass political stunt. This interpretation ignores the fact that, as slavery had been granted constitutional protection under the law of the land, the only open path to emancipation was by means of an amendment to the Constitution. No presidential edict or proclamation could free anyone—and withstand legal challenge. Yet Lincoln issued an Emancipation Proclamation. How did he do that, by what authority, and what did it mean?

As president of the United States, Abraham Lincoln had both executive and military powers, as do all presidents. As the executive of the land, there was little Lincoln could do except defend the Constitution as it stood and propose new legislation designed to remedy the flaw regarding bondage. But early in the war the country was neither psychologically nor intellectually prepared for a war geared toward emancipation or an amendment designed to end slavery and Lincoln was well aware of that fact.

> As President, he was responsible to the entire country, which obliged him to move with extreme caution and care; and right now most of the country seemed steadfastly opposed to emancipation even as a war measure. Should he now violate his pledge to leave slavery alone as an institution and issue an emancipation proclamation, the consequences, he feared, would be calamitous. Emancipation would almost surely drive the loyal border states out of the Union, alienate Northern Democrats and destroy the bipartisan war coalition that began after Fort Sumter, and maybe even ignite a racial powder keg in the North. Then the Union really would be lost.[7]

So executive action on the legal status of slaves—either an amendment to the Constitution or a proclamation as later issued—appeared in the early days of the war to be out of the question. An amendment would never have passed Congress, and if by some stroke of magic it would have, it would no doubt have been voted down by the people, who were at the time unprepared for such a dramatic change. By law, that left Lincoln with but one option, and that was his war powers as commander in chief.

By 1862 there was ample case law to allow for the confiscation of military materiel by the armed forces of the United States in time of war or rebellion from areas that were in hot dispute—but only in those areas that were in active dispute, and that fact is of critical importance. Since the Dred Scott decision had emphatically declared slaves to be nothing more than physical property, Lincoln decided to utilize his powers as commander in chief to confiscate property in the Southern Confederacy that was of particular use to the forces of insurrection—the slaves, in this case. But, in accordance with case law, he could do this only in those regions that were actively engaged in open dispute. This is the reasoning behind the Emancipation Proclamation, and the reason it frees only the slaves in the states still held by the Southern

Confederacy and does nothing in the Northern states, border states, or those areas in the South already under Federal control. The Emancipation Proclamation was a carefully crafted military edict, not a sweeping act of social reform, which the president had no executive power to issue, painstakingly designed to comport with the president's standing, legal authority. It reads like a legal brief because it was carefully crafted, not to inspire but to withstand legal challenge.

In the summer of 1863 Lincoln defended his reasoning regarding the proclamation in a letter to Joseph Conkling, and we will let him speak for himself:

> You dislike the Emancipation Proclamation; and, perhaps, would have it retracted. You say it is unconstitutional — I think differently. I think the Constitution invests its commander-in-chief, with the law of war, in time of war. The most that can be said, if so much, is, that slaves are property. Is there — has there ever been — any question that by the law of war, property, both of enemies and friends, may be taken when needed? And is it not needed whenever taking it, helps us, or hurts the enemy? Armies, the world over, destroy enemies' property when they cannot use it; and even destroy their own to keep it from the enemy. Civilized belligerents do all in their power to help themselves, or hurt the enemy, except a few things regarded as barbarous or cruel.[8]

Here then is the rationale for the proclamation; a military expedient put to social use. Lincoln had no power to emancipate anyone, but he did have the military authority to confiscate property he considered of use to any belligerent, although only in those areas that were being currently contested. Thus he could not confiscate war materiel in New York, Kentucky, Maryland, or New Orleans, for that matter, but he could in those areas of the Confederacy still under Confederate control. This is why the proclamation reads as it does, a fact seemingly lost on many modern observers who criticize the document's limited scope.

Still, the proclamation's effect was both immediate and powerful. It changed the entire meaning of the war; indeed, its psychological effect on both sides of the Mason-Dixon Line was momentous. Stephen Oates recounts one incident:

> Beyond the White House, a black preacher named Henry M. Turner ran down Pennsylvania Avenue with a copy of the proclamation and tried to read it to an assembly of Negroes. Out of breath, he gave the proclamation to Mr. Hinton, who read it "with great force and earnestness," and the blacks broke into uninhibited demonstrations, shouted, clapped, and sang of jubilee while dogs barked at their sides. Presently an interracial crowd gathered in front of the White House and called for the President. When he appeared at the window and bowed to them, the blacks cheered Lincoln, cried out in ecstasy, and said if he would "come out of that place, they would hug him to death." Preacher

Turner exclaimed that "it was indeed a time of times," that "nothing like it will ever be seen again in this life."[9]

In December 1862 Lincoln went a bit further, suggesting in a speech to Congress the idea of compensated emancipation for those Southern states that would give up slavery anytime prior to the year 1900.[10] This idea, like his proclamation, was met by many with a general sense of scorn.

The Emancipation Proclamation was simply the first step in a process that would eventually lead to legal emancipation and citizenship for blacks, and was, in all probability, all that was legally available to Lincoln at the time. To say the least, reactions were mixed, and in the South naturally negative. Reflective of this attitude was the journal entry made by Catherine Ann Edmondston at her plantation home near Scotland Neck, North Carolina, on December 6, 1862, belittling any notion of emancipation, compensated or other.

> Mr. Lincoln's message is published in our papers today and a weaker, worse written, bungling attempt at a State Paper never disgraced the archives. "Compensated Emancipation" is his Hobby, and he rides it John Gilpin fashion—blindly, ignorantly. He proclaims it as ending the war, as the Chinese formally did their Lanterns in the face of an enemy, sounds it as they did their drums, Ging-Galls & Gongs, & expects, like them, that overcome by its sound of terror that the South will lay down their arms, return to the bosom of the Union, & permit themselves quietly to be despoiled of their property under the specious tittle of Compensated Emancipation. The logic by which he attempts to convince the Yankee mind that this hobby of his is to pay the war debt is particularly amusing. He gravely admits that "it is not so easy to pay *something* as it is to pay *nothing*" & then comforts them with the assurance that "it is easier to pay a *large* sum than it is to pay a *larger* one, and it is easier to pay any sum *when* we are able than it is *before* we are able." Most sapient, Mr Lincoln![11]

Unfortunately, Mrs. Edmondston, along with most of the South, was not buying the president's line, and reactions like hers ranged from astonishment to sarcasm to outrage. "We are," she continued, "that is the 'Loyal' portion of the South, to be paid in bonds for our slaves which are to be emancipated & live in Utopian felicity with their old masters. White labor is to rise, negro labor not to depreciate — in short we are all to live in the enjoyment of the most unbounded prosperity, North & South, pay our debts, especially the *War* debt of the North, if we will only accept Mr Lincoln's grand panacea, Compensated Emancipation."[12] Yet despite the belittling attitude assumed by many detractors like Mrs. Edmondston, the Emancipation Proclamation had put in motion a process that could logically end only with the acceptance of all blacks into the Union as equal citizens under an amended constitution. It had changed the meaning of the war from a local squabble into a moral crusade, and it had done so within the boundaries of law.

That moral crusade would be amplified yet further at Gettysburg, Pennsylvania, in the fall of 1863 in a speech many consider one of the finest ever delivered. At Gettysburg it took Lincoln only three minutes and a mere 272 words to change the meaning of the American democratic experiment, from one that had been restrictive and static, to one that was inclusive and evolving. In a few short paragraphs Lincoln captured the meaning of the war as an act of national transformation which reshaped the national consciousness. "Though he'd turned down other requests to speak (he couldn't spare the time), Lincoln accepted the Gettysburg invitation, because he thought it an appropriate setting to say something significant about the meaning of the war, to explain how Union armies were fighting not just to subdue a rebellion, but to save democracy and America's liberal institutions."[13]

Lincoln gave his speech at the dedication of the new National Cemetery, but he had not been invited as the principle speaker that day. That honor had actually fallen to Edward Everett, and as Gary Wills points out, "Everett was that rare thing, a scholar and Ivy-League diplomat who could hold a mass audience in thrall. His voice, diction, and gestures were successfully dramatic, and he always performed his carefully written text, no matter how long, from memory. Everett was the inevitable choice for Wills [the organizer of the consecration], the indispensable component in his scheme for the cemetery's consecration."[14] Everett took the stage first and spoke for over two hours. His speech was reportedly informative and well received. Then Lincoln stood and gave his three-minute oration. This is what he said:

> Four score and seven years ago our fathers brought forth on this continent, a new nation, conceived in Liberty, and dedicated to the proposition that all men are created equal.
>
> Now we are engaged in a great civil war, testing whether that nation, or any nation so conceived and so dedicated, can long endure. We are met on a great battle-field of that war. We have come to dedicate a portion of that field, as a final resting place for those who here gave their lives that that nation might live. It is altogether fitting and proper that we should do this.
>
> But, in a larger sense, we can not dedicate — we can not consecrate — we cannot hallow — this ground. The brave men, living and dead, who struggled here, have consecrated it, far above our poor power to add or detract. The world will little note, nor long remember what we say here, but it can never forget what they did here. It is for us the living, rather, to be dedicated here to the unfinished work which they who fought here have thus far so nobly advanced. It is rather for us to be here dedicated to the great task remaining before us — that from these honored dead we take increased devotion to that cause for which they gave the last full measure of devotion — that we here highly resolve that these dead shall not have died in vain — that this nation, under God, shall have a new birth of freedom — and that government of the people, by the people, for the people, shall not perish from the earth.[15]

As Gary Wills explains, in his Gettysburg Address Lincoln reshaped the very conception of American democratic government: "The Gettysburg Address has become an authoritative expression of the American spirit — as authoritative as the Declaration itself, and perhaps even more influential, since it determines how we read the Declaration. For most people now, the Declaration means what Lincoln told us it means, as a way of correcting the Constitution itself without overthrowing it. It is this correction of the spirit, this intellectual revolution, that makes attempts to go back beyond Lincoln to some earlier version so feckless."[16]

And what Abraham Lincoln told us at Gettysburg was that the notion that all men are created equal was the central and defining tenet of the original republic; that the Civil War was a testing ground for that proposition; and finally, that the outcome of the war would determine whether that form of American democracy could in fact survive. Lincoln took the war, the motives of both sides, the almost incomprehensible slaughter, and somehow made sense of it all.

Many in the crowd were moved by the message, but not everyone that day enjoyed what they heard. While *Harper's Weekly*, for instance, declared that "it was as simple and felicitous and earnest as was ever spoken,"[17] the *Chicago Times* drubbed the speech, declaring that "Mr. Lincoln did most foully traduce the motives of the men who were slain at Gettysburg,"[18] this in reference to his use of the term "a new birth of freedom." The *Springfield Republican* thought that "his little speech is a perfect gem; deep in feeling, compact in thought and expression, and tasteful and elegant in every word and comma,"[19] while the *Times* of London clearly thought otherwise, suggesting that "the ceremony was rendered ludicrous by some of the sallies of that poor President Lincoln.... Anything more dull and commonplace it would not be easy to produce."[20]

By and large, the response to Lincoln's address was positive, however. But by far and away the kindest and most insightful comment came from Edward Everett himself, who sent a short note to the president after the ceremony stating, "I should be glad if I could flatter myself that I came as near to the central idea of the occasion, in two hours, as you did in two minutes."[21] As Gary Wills suggests, over time the Gettysburg Address seeped into the national consciousness, and Lincoln's explanation of the war and interpretation of the national mission would be accepted by most Americans as valid: a war that was moral and a mission that was inclusive.

In early March 1865, with the Civil War drawing toward its inevitable conclusion (Lee would surrender his army on April 9, 1865), Abraham Lincoln again addressed the nation, this time at his second inauguration. This speech is recalled most principally for its last paragraph, a paragraph that is so radically different from the victory speech and victory documents of prior wars and prior epochs that it is almost astonishing. There is no boasting, no swagger, no self-validation, no demeaning of the soon-to-be defeated enemy, no thumping of the chest. This closing paragraph, rather, brims with care and compassion, not simply for those who fought for the North, but for *all* combatants, and for all citizens. Failing to distinguish between victor and vanquished, it enjoins the entire nation, North and South alike, to move forward with consideration and kindness toward one another. It strikes a remarkable tone, and, truly, there is nothing quite like it in the war-tattered history of humankind prior to this moment.

Historically, wars did not generally end well for those who found themselves on the losing side, with immediate death or a dark prison cell the most likely prospect, and most definitely they did not end well for those in rebellion. Spartacus, as an example, defied Rome, and when he was defeated his entire following was crucified upside down, from the gates of Rome to the sea, a gruesome warning to anyone contemplating revolution against the Roman Empire. As a reminder of the sort of victory pronouncements prior levels generally issued — and in stark contrast to Lincoln's inaugural speech — recall the triumphant boasts of Ashurnasir-pal after sacking a defiant town: "I cut off their heads. I burned them with fire, a pile of living men and heads over against the city gate I set up. Men I impaled on stakes. The city I destroyed, devastated. I turned it into mounds and ruin heaps, the young men and maidens in the fire I burned."[22]

Now Mr. Lincoln speaks: "With malice toward none; with charity for all; with firmness in the right, as God gives us to see the right, let us strive on to finish the work we are in; to bind up the nation's wounds; to care for him who shall have borne the battle, and for his widow, and his orphan — to do all which may achieve and cherish a just, and lasting peace, among ourselves, and with all nations."[23] Here, truly, are on display the compassionate sentiments of the level four/five mind, not justice and compassion only for the individual clan, tribe, city, or nation to which one is a member, but justice and compassion for *all*. Indeed, so radically different are Lincoln's sentiments from those of Ashurnasir-pal, the two men seem almost beings of a different species.

Those high, level four/five sentiments are what drove Abraham Lincoln through the course of the Civil War, and what, by late January 1865, would produce the successful passage of the 13th Amendment to the Constitution, which reads simply enough: "Neither slavery nor involuntary servitude, except as a punishment for crime whereof the party shall have been duly convicted, shall exist within the United States, or any place subject to their jurisdiction."[24] With that the great stain that had haunted the republic since even before its creation had been expunged, slavery abolished, and a "new birth of freedom" made a reality. Stephen Oates captures the excitement of the moment:

> The vote came on January 31, 1865, with spectators filling the corridors, and galleries of the Capitol and Lincoln awaiting the outcome in the White House. As the roll call proceeded, every Republican member of the House voted yes, until at last the chamber hushed and the clerk read out the results: 119 for the amendment and 58 against — just three votes more than the required two-thirds majority. At once "a storm of cheers" broke out among the Republicans, who jumped around, embraced one another, and waved their hats overhead. There were shouts and applause in the galleries, too, where women's handkerchiefs were fluttering in the air. In all the rejoicing, Representative George Julian of Indiana said he felt as though he were in a new country.[25]

The Emancipation Proclamation, the Gettysburg Address, and Lincoln's second inaugural address all clearly embody the core values of level four thought: democracy, compassion, fairness, tolerance, equality, mercy, inclusiveness, and others. As documents of war they rectified what had been ignored during and after the Revolution and set the United States on a new course toward a future that more accurately reflected in fact what before had been in theory only, thus unifying the nation as never before. "Up to the Civil War," observes Gary Wills, "the United States was invariably a plural noun: 'The United States are a free government.' After Gettysburg, it became a singular: 'The United States is a free government.' This was the result of the whole mode of thinking Lincoln expressed in his acts as well as his words, making *union* not a mystical hope but a constitutional reality."[26]

Abraham Lincoln looked at slaves and saw human beings, whereas many in the South — and in the North, truth be told — saw only property. It was the difference between the Dred Scott decision and the 13th Amendment, between the world as seen through the eyes of levels two and three, and the world seen through those of level four. Most people today cannot conceive of a time when people looked upon others as mere property or worse. Unfortunately, that view had been the way of the world for most of the human epoch, and only recently has it begun to change. I refer again to Dr. Clifford Anderson: "Maturationally speaking, our species has had a busy past century. During this time the populations of most of the advanced countries of the world have evolved from a general level of maturation that was arrested in childhood to the point at which — for the first time — many are knocking on the door of psychological maturity."[27] Frankly, it requires no great effort to peer into the historic record to see that this was so, indeed, that it was the *reason* the Civil War became such a long and bitter conflict; and across our world today that clash continues to be played out time and time again.

In four years — an amazingly short period of time when all the obstacles are considered — Abraham Lincoln (along with a few million other men and women) rectified the moral hole in the Constitution the founders had refused to touch, that had stymied, and harried, and flummoxed the country for the preceding eighty years. And he paid for that success with his life. "*Sic semper tyrannus*"[28] (thus always with tyrants), John Wilkes Booth was supposed to have shouted after assassinating Lincoln at Ford's Theater in Washington, a declaration that could only have made sense to a level three mind that easily considered human beings as property; thus to that mind he who would free them was a tyrant. It was, of course, the last dying shriek of a fated way of life, and Booth would go down in infamy while Lincoln is today regarded as one of the finest American presidents ever to have lived. Yet the way of life Booth extolled was still alive and well on the European continent, and soon it too would have its final death dance in a catastrophe known as World War I.

Twelve

A War to End All Wars

The same dynamic forces that exploded in the American Civil War would collide again on the European continent some fifty years later, and with frightful results. And just as with the American Civil War, very few people would grasp the horror that was coming.

From our historical vantage point it is easy now to tease those dynamic forces apart and examine them individually; but for those living at the turn of the nineteenth century, they were either functions of the way things had always been, or else relatively new phenomena that had yet to be given serious consideration. Yet these forces would soon produce a conflagration that simply dwarfed all prior conflicts. The Great War, it would be called, because, for those who managed to survive those four awful years, no additional description seemed necessary.

An ocean away the American Civil War had brought about the utter destruction of the level three plantation South on the North American continent and truly unified the United States as never before. From that moment forward, the national agenda would revolve essentially around level four issues, and while it would take almost one hundred years to overcome stiff level three attitudes among the body politic, basic civil rights would finally be accorded all citizens regardless of race, gender, creed, or color. As the world raced ahead into the twentieth century, the United States was rapidly fulfilling its original revolutionary promise.

The United States had paid a terrible price for the incorporation of slavery within its initial composition, but the Old South was at least now a thing of the past and a new South fast arising from the ashes. Unfortunately, the rest of the world was still clinging to the monarchial remnants of a level three ruling structure, and unwittingly drifting toward a series of conflicts that would blast away the last of those remnants while making the killing fields of the American Civil War look like little more than child's play.

The twentieth century dawned on Europe much as had centuries past. Europe was a continent ruled by dynastic families that in some cases stretched back for hundreds of years. In Russia the Romanov Dynasty still ruled over a vast and backward land, while in nearby Austria-Hungary the Hapsburgs clung tenaciously to the remnants of their once-great empire. The Ottoman Empire still stretched from Bosnia in southern Europe all the way to Egypt, but it was a corrupt and teetering shell of its former glory. Ruled by the house of Hohenzollern in central Europe, Germany — the last of the great powers to aspire to modern nationhood — had only a few decades before been pounded together by Prussian predatory efforts.

While an entirely new century had dawned, no one in power on the European continent appeared to perceive the world differently than they had for the last five hundred years. "As the world entered the twentieth century, it carried with it a host of dynasties who regarded their right to govern as a divine dispensation."[1] In a very real sense, the European kings, czars, and kaisers had inherited a worldview not terribly different from the one Alexander the Great had inherited from his father, Philip. To them, notions such as democracy and equal rights appeared to be no more than passing, although dangerous, fancies, and the United States little more than a curious anomaly. The way of monarchy had been the way of the world for centuries, and these leaders sensed nothing on the wind that might affect their divine right to rule.

Yet Europe *had* changed, and in many ways radically. If that change was not yet reflected in the worldview of the European leadership, it was most certainly reflected in the anarchy that was rattling throughout the working classes from Constantinople to Moscow, the strident nationalism that was simmering in the Balkans, and the technology that now blanketed much of the continent. Level four consciousness had not yet arrived in full, but it was most certainly knocking at the door.

European armies, although essentially at peace since the Franco-Prussian War of 1871, had changed dramatically. Not only had weaponry been vastly improved since the end of the American Civil War, but the technological capacity to feed, move, and support enormous armies in the field now existed throughout most of France, England, and Germany.

At the turn of the century Germany, as an example, maintained a ready force of regular and reserve units of some 2.2 million men.[2] Compare that with the meager American standing army of only 16,376 officers and men at the start of the Civil War only thirty-nine years earlier, and you will get some idea as to the enormous leap in forces available at the start of this new century. And the infrastructure needed to support these armies was already in place.

The rail systems, for instance, that hauled millions of workers, travelers, and tourists daily could, in the blink of an eye, be equipped to move millions of soldiers to and from the front and supply them with food and ammunition.

The field telegraph, in its infancy during the Civil War, had now been replaced by field telephones, an evolution that would greatly augment battlefield communications. Field artillery was now enormously more accurate and lethal, and the 1861 single-shot Springfield had been replaced by machine guns with names like Maxim, Browning, and Madsen. So while the ruling autocrats and their military staffs remained devoted to a worldview that was rapidly crumbling, they were in their divinely inspired somnolence unwittingly setting the stage for a calamity of almost unimaginable horror, and a calamity that would ultimately eventuate their own demise.

Lurking just offstage in all of this, but soon to be the principle player in the production, were German territorial aspirations. United under the iron hands of both Frederick the Great and, later, Bismarck, Prussia, the center and driving force behind the German nation, had often been called little more than an army masquerading as a nation. While Germany was now united, the Prussian warrior worldview still dominated both the military and civilian hierarchy. Perhaps not since the days of Sparta had a state been so militarily focused, and by 1914 that focus was glancing enviously across the European landscape.

"No German could forget that short, brilliantly planned wars such as the Franco-Prussian conflict of 1870 had been the means whereby their nation had been expanded and united, and Germany's rulers fully expected to use the same methods to bring the whole of Europe under German leadership."[3] Germany was in fact only on the verge of being a modern state, and because of its Prussian, warrior ethos, it remained very much a predatory, level three society.

Robert O'Connell captures this German dichotomy with precision, noting that while German educational and governmental systems were modern in form, the nation's myths and imagination were largely involved in a fictitious past.[4] The German military, armed already with an offensive blueprint for conquering France called the Schlieffen Plan, sat waiting and watching only for an opportunity to pounce.[5] They would not have long to wait.

In *The Guns of August* historian Barbara Tuchman describes the evolving, German, classic level three predatory ambition with rare insight:

> A hundred years of German philosophy went into the making of this decision in which the seed of self-destruction lay embedded, waiting for its hour. The voice was Schlieffen's, but the hand was the hand of Fichte who saw the German people chosen by Providence to occupy the supreme place in the history of the universe, of Hegel who saw them leading the world to a glorious destiny of compulsory *Kultur*, of Nietzsche who told them that Supermen were above ordinary controls, of Treitschke who set the increase of power as the highest moral duty of the state, of the whole German people, who called their temporal ruler the "All-Highest."[6]

Just as with the American Civil War, all that was needed was a spark, something that would put events into motion, and that spark would be supplied on June 28, 1914, when Serbian anarchists assassinated Archduke Francis Ferdinand, the heir to the Austro-Hungarian throne, along with his wife, while the two were on a visit to Bosnia. When the conspirators were arrested and it was determined they had been trained and equipped by the Serbs, Austria sent an ultimatum to Serbia that was so severe in its demands the Serbian rulers had little choice but to refuse to comply.

So Austria declared war against Serbia, then began to mobilize its troops. Russia, a Serbian ally, began to prepare for war as well. Germany, now fearing a possible war on two fronts (potentially against Russia on the east and France on the west) promptly declared war on Russia.

But Germany's Schlieffen Plan had been designed for war against France by way of Luxembourg and Belgium, two neutral nations, not Russia. So, trumping up false and ridiculous charges of aggression, the Germans declared war against France on August 3, and the next day began their much-coveted invasion by way of Belgium. Outraged by the invasion of a neutral, Britain added the last piece to the puzzle by declaring war against Germany. The die had been cast. In just slightly more than a month Europe had gone from what appeared to be a peaceful slumber to total war. When war began, few nations beyond Germany had any conception of where it would lead or, for that matter, even a plan for immediate application. Indeed, the war's genesis seemed entirely irrational, but we must remember that predatory war is not a rational undertaking in the first place, no matter how much scheming or pre-planning has taken place.

Predatory warfare is psychologically motivated — a grossly mistaken reaction to fear that is projected outward. Alexander and his numerous kindred spirits had for thousands of years responded to this powerful emotion by means of predatory conquest. This motivation, in the hands of the warrior class for centuries, had by 1914 been reduced to the deepest of ruts, and thus the decision-making options available seemingly narrow indeed. Psychological reactions had been reduced to narrow behavioral patterns which were being repeated again and again with precious little thought. To the German military hierarchy, war was simply a way of life and conquest a natural course that required little if any justification. Stanton Coblentz gets it exactly right: "To find more basic reasons [for war], the observer would have to peer into the mind of a hundred warring generations, and the code of slaughter and destruction they had formulated and strengthened."[7] To the autocratic mind of the early twentieth century war simply called for more war, and thus the world stumbled headfirst into a tragedy the dimensions for which it was entirely unprepared.

Rather than gloom, however, all across Europe news of war was greeted with an initial sense of almost euphoric liberation. "The coming of war," notes John Keegan, "was greeted with enormous popular enthusiasm in the capitals of all combatant countries."[8] Eerily reminiscent of the American Civil War, most everyone thought the troops would be home by Christmas, only four months away.

The Germans quickly drove through Luxembourg and Belgium, then on into France, initiating as they did the systematic horrors and slaughter of innocents for which German arms would become increasingly associated over the next four decades. "Within the first three weeks," Keegan explains, "there would be large-scale massacres of civilians in small Belgian towns. At Andenne there were 211 dead, at Tamines 384, at Dinant 612. Worst of all the outrages began on August 25 at Louvain. This little university town was a treasure store of Flemish Gothic and Renaissance architecture. At the end of three days of incendiarism and looting, the library of 230,000 books had been burnt out and 1,100 other buildings destroyed."[9]

In particular the Germans became upset and obsessed with any disorganized resistance (the *franc-tireurs,* or civilian snipers) they routinely encountered, a natural response for a free people, yet a response entirely incomprehensible to a level three, predatory culture. Again, Barbara Tuchman comments: "Schooled in a state in which the relation of the subject to the sovereign has no basis other than obedience, he is unable to understand a state organized upon any other foundation, and when he enters one is inspired by an intense uneasiness. Comfortable only in the presence of authority, he regards the civilian sniper as something particularly sinister."[10] And responds, we might add, accordingly.

Bashing, burning and murdering their way through Belgium, the Germans swept into Northern France just slightly behind Schlieffen's timetable for conquest. There the French military, adrift in a level three Napoleonic delusion all their own, were waiting for them as if mustered for battle in the previous century. "The infantry wore bright blue coats and scarlet trousers, and the cavalry was resplendent in gleaming breastplates," Tuchman tells us.[11] The French sense of spirit, or élan, had for decades vigorously resisted any change to a more practical field uniform of perhaps brown or grey that would be less conspicuous on the modern battlefield. At a hearing in parliament called to discuss the elimination of their famous red trousers, for instance, M. Etienne, fumed, "Never! *Le pantalon rouge c'est la France!*"[12] Thus did the French attack grandly only to be slaughtered grandly. In only two weeks the French lost over 300,000 men and France itself appeared doomed, scarlet trousers and all.[13]

But by early September the Germans were weary and shot-up too.

Wheeling north of Paris in an attempt to bag the staggering French army in a double envelopment, the exposed right flank of the German right wing was spotted and vigorously attacked. Within days the Germans were forced to withdraw back beyond the Marne River and dig in. The French attacked aggressively, but frontal infantry assaults proved so disastrous that for a period of weeks each side tried to do nothing more than progressively outflank one another. The result was a front that was daily stretched and that ultimately extended from Noyon, on through Flanders outward to the English Channel. This exercise in futility eventually left two parallel lines stretching across Europe from the North Sea clear to Switzerland.[14] The battle lines now stretched for almost 500 miles, the two combatants facing each other across a "no man's land" raked by machine-gun, artillery, and rifle fire and where life was often measured in little more than seconds.

Here the Great War would take on its most distinctive and enduring characteristic: bitter, bloody, fruitless trench warfare. And the reason, which no one — just as during the American Civil War — seemed to grasp at the time, was simply the extraordinary increase in lethality of small arms and field artillery. In particular the machine gun had made trench defenses almost impenetrable, because a machine gun is in the final analysis just that — a machine. Once erected, it simply mows down everything in its path with frightening precision.[15] Positioned at the extreme ends of a trench system, the overlapping, enfilading fire of machine guns created a kill zone where virtually nothing advancing into the zone could survive. Yet large units of men were time and again fed into the teeth of these defenses no matter how utterly futile the results, or horrific the slaughter.

Initially both combatants craved a war of maneuver, in which speed and striking power would prove supreme. But the Great War disappointed, settling into something akin to an enormous and ungovernable siege, vastly limiting even the hope of maneuver. This fact seemed lost on the commanders of both sides, however, who systematically attempted breakthrough after breakthrough in order to gain the open ground necessary for maneuver; but they failed with an appalling consistency, sacrificing thousands upon thousands of men's lives for their delusions of tactical maneuverability.

In essence the military planners had entered the Great War prepared to fight not the opposing armies as they actually existed or the situation on the field, but rather an opponent from the past, something far more akin to Lee's Army of Northern Virginia. Speed and maneuver were still thought to be the keys to victory. Thus the opposing combatants tried desperately to break the maze of trenches and gain the open field by means of horrendous frontal assaults, all of which failed, often miserably. For Lee's army was long gone, and it would take the next four years to finally figure out that massive infantry assaults against well fortified, entrenched positions were nothing more than invitations to mass slaughter. By the outbreak of World War I, developments in small arms ordnance and field artillery had made the war of maneuver that both sides desired physically impossible to fight.[16] Unfortunately that stopped no one from fighting. The weapons men had created in the vain hope of producing a tactical imbalance on the field of combat had in fact made war as it was then conceptualized almost impossible to fight. Weapons, not men, had taken command of the field.

The First World War would drag on for a horrible four years, and Germany would find itself fighting a war on two fronts, the very thing its leaders had feared at the outset. Level four military technology had simply far outdistanced the level three Homeric codes of war that still favored large, frontal infantry assaults to carry the day. The urge to close remained paramount in the minds of military planners. Just as in the American Civil War, this painfully deficient tactic would go essentially unrecognized by the military leaders on both sides of the conflict as men were continually fed into the whirlwind like meat into a grinder. Victories were often measured in nothing more than a few hundred yards gained of treeless, cratered, battle-scarred mud. "The most carefully prepared war plans were reduced, sometimes in a matter of moments, to antic apologies for slaughter."[17]

Somewhere close to ten million men would perish in combat operations during World War I, a loss of almost an entire generation of European males — and all for nothing of substance.[18] J.F.C. Fuller notes that for over a period of four years, no offensive by either side on the Western Front was able to shift the front line even ten miles in either direction.[19] The industrial revolution, offspring of the rise of level four consciousness in the West had, in less than a century, managed to create weapons so lethal that they had reduced Homer's heroic battlefield to little more than a slaughterhouse.

Ultimately this mindless slaughter would lead not simply to a sense of general demoralization but to an abrupt and profound sense of cultural shock. Not only was the war questioned, but the entire course of human history that had somehow led to it fell into doubt. Prior to this moment war had been viewed as essentially heroic: never again. Robert O'Connell notes that the war's ghastly outcome "was such a shock, and raised such profound questions about the basic directions of Western civilization, that it created a crisis of the spirit unparalleled in modern times."[20]

Scientific technology, a single thread in the multifaceted emergence of level four consciousness, had made war, as it had been conceived and fought for over twelve thousand years, essentially obsolete. The ancient Homeric traditions had become suddenly outmoded. The martial qualities that had set apart the finest warriors and had become the stock-in-trade of the Homeric ethos were suddenly rendered impotent. Men were blown apart by distant artillery or mowed down by unseen machine guns long before they had a chance to close with the enemy; "there was hardly a heroic death to be had."[21]

The Great War would end, not in a breakthrough victory by one side or the other but in the futility of an armistice called with all combatants essentially in the same positions they had occupied for the previous four years. In that sense the war did not end so much as play itself out, allowing the survivors to make of it what they pleased.

For the French, English, and Americans (who were ultimately drawn into the conflagration in 1918) the armistice was interpreted as a great victory, but the Germans remained equally steadfast in their belief that they had not, in fact, been beaten. Thus Germany's cultural myths, which had initiated the horror to begin with, remained essentially intact, ready to flower again if only properly cultivated, which soon they would be.

Still, the Great War marked a watershed event in human affairs. From that moment forward the emerging level four mind would only rarely equate war with either glory or heroism. To the contrary, in fact, a pacifist movement would slowly take hold throughout the United States and much of Europe that would equate war, and its concomitant Homeric warrior ethos, with abject failure, rank stupidity, and gross cowardice. Soon pacifism, a profoundly level four and emerging level five (a level that will be discussed in subsequent chapters) moral imperative, would claim for itself the Homeric quality of heroism. Suddenly to *not* fight was for some few deemed far more heroic than to fight. Clearly the world was in the throes of an enormous psychological and intellectual transformation.

But pacifism, born of level four/five idealism (the ability to conceive what might be) would misinterpret the world almost as profoundly as had the level three militarists it so stridently abhorred, and thus it would continually fumble away its strong moral insight. Denouncing the industrial West, it would attack the values of the level from which it had sprung (level four, the level that actually nurtured it into being) while failing utterly to distinguish between level three predatory warfare — which was, in fact, the problem — and the entirely justifiable level four defensive response. Without understanding human development it is impossible to understand war, and thus pacifists, with a rising sense of moral superiority, simply associated *all* war and *all* things military with abject failure, rank stupidity, and gross cowardice. This was a mistake.

What pacifists failed to see at the time, and continually fail to recognize to this day, is that level four, the very level that gave pacifism life to begin with, would have long ago, without reasonably stout defensive mechanisms, been rapidly consumed by predatory, level three aggression. Recall, for instance,

Twelve. A War to End All Wars

Infantry advance into "No-man's-land" during World War I, a rendition by Lucien Jonas; this tactic would prove virtually suicidal due to advances in the volume and coordination of artillery and machine-gun fire (Library of Congress).

Finland's bloody encounter with the Soviet Union. Democratic institutions did not emerge like lilies in the field. They were fought for, often long and bitterly, and pacifism emerged because of that fight, not despite it. Pacifism, abhorring all war, generally fails to detect the genuine moral objectives inherent in most level four struggles. Slaves in the American South, for instance, were not freed nor were the gates to the Nazi death camps in World War II thrown open by meditators, evangelicals, or pacifists. These tasks were accomplished by vast contingents of level four soldiers, many of whom gave their lives to see it so.

Pacifism *is* a high moral calling of level four/five development, but that high calling must be tempered with practical reality, something pacifists often vigorously resist. But that is hardly surprising. As we have seen, all developmental levels arrive in their most immature form, only then to mature and adjust over time; no doubt pacifism will one day mature to a point where it can both embrace and be embraced by a majority of humanity. That would be a wonderful and profound development.

But the world's population must first develop to the point that democracy — the first spontaneous counterbalance to impulsive, predatory warfare — can arise naturally across the globe. Unfortunately, that remains a Herculean task. As Ken Wilber points out, even today some 70 percent of the world's population remains below level four, developmentally incapable of sustaining a democratic form of government, much less a pacifist worldview.[22] So before we can seriously talk of ending war through mutual, informed agreement, it may be wise to begin paving roads, planting crops, building schools, and, most important, caring for children — in other words, creating the requisite *conditions* that induce level four growth, which will in turn someday naturally evolve into democracy, and a much safer world for us all.

Trench system utilized during World War I on the Western Front. Troops had been driven into the trenches due to the sheer firepower of artillery and small arms ordnance; improvements in weaponry would make the breakout war of maneuver desired by both sides a martial fantasy during most of the First World War (Library of Congress).

World War I cavalry on patrol. While the cavalry was still in use as late as World War II, the lethality of small arms ordnance had, by the time of the outbreak of the First World War, reduced its role to reconnaissance and supply line protection (George Grantham Bain Collection).

The Great War would go a long way toward removing the old monarchial (level three) structures that had impeded development for centuries. "Following a revolution in 1917, the vast expanses of imperial Russia emerged as the Union of Soviet Socialist Republics, the world's first Communist state. Germany lost its emperor and became, briefly, a republic. Rent by internal dissent,

the Austro-Hungarian empire dissolved into its component parts. And in the Middle East, the once-mighty Ottoman Empire lay in fragments."[23]

The most creative proposals that would appear either during or after the war were Woodrow Wilson's Fourteen Points, the central feature of which was a League of Nations, an organization much in the mold of the modern United Nations. Outlining them during a speech to a joint session of Congress on January 8, 1919, some few months before the armistice, Wilson, voicing recommendations aimed at furthering world peace — and in terms that Ken Wilber would today term world-centric — made a strong appeal for justice and democracy:

> What we [the United States] demand in this war, therefore, is nothing peculiar to ourselves. It is that the world be made fit and safe to live in; and particularly that it be made safe for every peace-loving nation which, like our own, wishes to live its own life, determine its own institutions, be assured of justice and fair dealing by the other peoples of the world as against force and selfish aggression. All the peoples of the world are in effect partners in this interest, and for our own part we see very clearly that unless justice be done to others it will not be done to us.[24]

Wilson's Fourteen Points, most of which addressed the just distribution of war-affected territories, were far reaching and crafted from level four/five patterns of thought; but they were in the end simply too far ahead of their time to find much in the way of an appreciative audience and subsequently found little resonance either at home or abroad. It would take another thirty years and another world war before much of Wilson's thinking would be both appreciated and put into practice, most principally in the form of the United Nations. It should be noted, however, that deliberative bodies such as the American Congress, the British Parliament, and the United Nations, as examples, are in fact level four democratic institutions; and democratic institutions, in the hands of undemocratic members (level three), can be destroyed, compromised, or bent to predatory goals. This is the reason the United Nations' record thus far has been fair but far less than spectacular. The problem arises not from the institution so much as the membership, but it is to be hoped that will change for the better in the coming decades as human development continues apace.

Opposite, top: **British artillery in action. Artillery, in conjunction with forward observers, would come to dominate the World War I battlefield, rendering time-honored infantry tactics obsolete (Library of Congress).** *Bottom:* **The tank (British), introduced during World War I initially, like the ancient battle elephant was successful as much for the shock and fear it produced as its battlefield capabilities. Named tanks by the British to fool the Germans into thinking the units were mechanized water transports, the military tank would soon come to dominate the field of battle (National Photo Company Collection).**

Unfortunately, after World War I a majority of the undemocratic European population and its political leadership still hovered around level three, thus confusion and turbulence rather than democracy and pacifism were quickly ushered in. Europe, expanding rapidly beyond level three, but not yet at level four, for the few decades following the Great War would remain in a sort of developmental limbo. A worldwide depression along with the reparations Germany had been saddled with at the Treaty of Versailles which ended the Great War would bring the new German republic to its financial knees. Political, social, and economic instability ruled the day.

Stepping directly into the center of this combustible mix would come a new German leader with an agenda designed to raise triumphant the old Germanic mythic delusions of national and racial superiority. Adolf Hitler had arrived, and the world would never again be the same.

Thirteen

The Great War, Act II

What came next had the odd, recurrent feel, as the American ballplayer Yogi Berra once put it, of "deja vu all over again."

The Germans, with an eye toward carving up Russia, but not wanting to get caught in the middle of a war on two fronts, decided to first invade France before turning back to consume the Soviet Union. So the Germans raced through neutral Belgium, taking everyone by surprise and shocking the world. French forces, unprepared and caught off guard, were instantly overwhelmed by the German advance, then were forced to fall back on Paris in rout and confusion. But the date was not August 1914. Rather it was some twenty-six years later, May 1940. The curtain had been drawn, it might be said, on the Great War, Act II.

The same misbegotten notions of national and racial superiority that had ignited the First World War had resurfaced in Germany, once again driving Europe to the brink. World War II was in large measure simply World War I reignited. "Its roots," Robert O'Connell observes," are to be found tangled in the conflict which preceded it, not merely in terms of causation but also in the very manner in which it was conducted."[1]

The principal difference between the two wars, of course, was that during the Second World War German armies would no longer be under the direction of the kaiser and his general staff but rather the pathologically distorted mind of Adolf Hitler. That one difference would change the second act of the Great War into an orgy of blood and inhumanity that even today remains almost beyond comprehension.

As demonstrated in the preceding chapter, the First World War had not ended in 1918 so much as it had collapsed from exhaustion. Both the will and the means to conduct the war had been bled dry throughout most of Europe, ultimately leaving the combatants utterly spent while resolving nothing of substance. Thus German bitterness and racial contempt would continue seething just below the surface for decades while the rest of Europe went about the hopeful, if somewhat quixotic, pursuit of peace and disarmament.

Robert O'Connell, with no eye toward developmental theory, interprets the in-between war years with remarkable insight, noting that a genuine desire for peace animated most European populations, but not necessarily their governments: "The difference lay in governmental systems and their relationships to the respective military establishments. In the democracies [level 4], the pacifistic inclinations of their populations was faithfully reflected in official policy, while in the totalitarian states [level 3] the military agenda counted for much more and could be pursued with equivalent rigor."[2]

Germany, of course, had a brief fling with democracy during this period, but economic instability and deep-seated national resentments allowed Hitler — just as with Mussolini in Italy — to bully, murder, and bludgeon the republic into submission. The world was still numb as a result of the Great War, and in no condition to resist such a fanatical, hate-mongering zealot. "It was his fate to seize the reins of a sick world, unwell physically and mentally as a result of a wrenching four years' conflict, a world still in large part unspeakably war-weary and eager for peace."[3]

Hitler's methods were focused, methodical, and brutal. "Democracy," he declared, "must be destroyed by the weapons of democracy."[4] No scheme or violence was beyond or beneath him. Thus "there was the political trickery, the crimes such as the burning of the Reichstag building and the liquidations of intellectuals and anti–Nazis; there was the 'blood purge' of 1934, by which an unknown number of Hitler's collaborators were unofficially put out of the way."[5]

In short, Adolf Hitler used democracy's finest characteristics to first destabilize then destroy the German republic, and the ease with which both Hitler and Mussolini achieved their parallel goals should give pause to anyone firm in the belief that democratic institutions can flourish unaided in undemocratic soil. As pointed out in the previous chapter, democratic institutions are only as good as the people who employ them; and in the hands of those who have no respect for freedom's goals, they will either wither and die, or be bent to destructive ends. Democratic institutions require democratic minds to understand, run and respect them, something level three minds, by definition, are generally incapable of.

By 1933 Hitler had bludgeoned his way to the German chancellorship, but nothing short of complete power would satisfy him. "In a parliamentary chamber dominated by a huge swastika and ranks of SA troops, the frightened deputies granted him dictatorial powers by a large majority. (They had good reason to be frightened: Of the ninety-four dissenters, twenty-four were subsequently murdered.)"[6] Hitler had forced his way to complete power, and by 1933 the infant German republic lay in ruins. "After the death of German President Hindenburg on August 2, 1934, the offices of president and chancellor were combined, and Hitler became the furher, the ultimate leader. Like Stalin, he achieved nearly complete control over the state."[7]

Adolf Hitler's plans for Europe, previously published for the world's edification in his book *Mein Kampf* (My Struggle), were of the most predatory imaginable.

> In the Soviet Union, Germany would find the lebensraum, the living space, its people needed.... The Soviet Union was to be utterly crushed, its cities leveled, it people enslaved. In this new empire, the Jewish "problem" would find its "final solution" in the total eradication of Jewish culture.... It was a vision devoid of humanity, spirituality, or creativity; it was a barbaric vision, deliberately so, for it was through barbarism, Hitler believed, that the dynamic, healthy new culture would replace the degenerate old one.[8]

O'Connell captures Hitler's bleak, violent, predatory character with unblinking clarity: "His true milieu was war of the most predatory sort. Those who wonder what might have happened had Hitler not resorted to force in 1939 or not invaded Russia in 1941 are dealing on the level of fantasy."[9]

But it was really far more than mere conquest that drove Hitler to the violent extremes employed by his German Reich. He was consumed by hate, and death-dealing seemed his only natural outlet and expression. "Death was the leitmotif of his regime — so much so that a major governmental function became the hunting and ritual slaughter of millions of noncombatants deemed subhuman, the Jews.... This was no minor preoccupation."[10] Not minor at all. In the end, it is estimated that the German Reich executed over six million Jews, and an additional twelve million Russians, Poles, Ukrainians, Belorussians, Gypsies, and Serbs.[11]

Death-dealing on a level never before conceived was Hitler's true contribution to the world, one humanity has been trying to come to grips with ever since. And it seemed not to matter too terribly much to Hitler just who was consumed in the effort, himself and all his henchmen included. As early as 1932, for instance, he was quoted as telling Raushning, "We may be destroyed, but if we are, we shall drag a world with us — a world in flames."[12]

In that sense, Hitler's plans, schemes, and strategies can all be viewed as a death wish as much as they can fantasies of conquest, and a "world in flames" his last will and testament. Hitler, in that sense, is our serial killer from chapter five, only now with the murderous capacity of a modern military and a cowed, willing nation at his fingertips. The result was not simply war. It was murder, slaughter, extermination on an unimaginable scale. "It may not be without significance that, having perpetrated this greatest of historical atrocities, the Germans should have given us a new word (one officially recognized by the United Nations), the word 'genocide.'"[13]

The road to lebensraum began in March 1936 when columns of German troops reoccupied the Rhineland, in violation of the Treaty of Versailles, and the world simply looked the other way. At the time, the occupation was viewed as little more than a minor irritant by an international community weary of war. Then in 1938 Hitler's troops marched into Austria. The French did nothing. The British responded with a policy of "appeasement," by which it was hoped a sensible and lasting peace could be crafted. "At the time the policy was seen as a noble ideal: magnanimous, sound security, good business."[14] Today, of course, the appeasement of predatory dictators is viewed as the very height of folly because appeasement does not satisfy the pathological urge for conquest but rather feeds and sustains it.

The leading advocate of appeasement at the time was British prime minister Neville Chamberlain. Today Chamberlain is viewed as both a fool and a dupe, yet those criticisms, upon consideration, appear at least somewhat unfair. Chamberlain did nothing more than accurately reflect the peaceful intentions of the world's democracies, and presumed — quite incorrectly, history would prove — that he was dealing with a German regime that shared this same democratic longing for peace. "It never occurred to Chamberlain that Hitler did not want peace, that his whole personality, the personality of the Nazi regime, demanded the waging of war."[15]

In essence these were negotiations taking place between the advanced level four/five mind — war-weary, pacifistic, genuinely hoping to create a new and safer world — and the most extreme example of predatory, level three aggression the world had ever produced. And the bumbling folly of these negotiations serves as an interesting example of all that can go wrong during the course of inter-level discussions. Both levels came to the table essentially incapable of understanding one another. Chamberlain, thoroughly imbued with level four/five high-mindedness, believed Hitler to be a man like himself, of good character and desirous of peace. Hitler saw in Chamberlain's willingness to compromise and appease only weakness, and, as we have seen, level three has nothing but contempt for weakness. Both men were wildly mistaken. Chamberlain vastly misjudged Hitler's motives, while Hitler vastly misjudged the willingness of democratic nations to stand and fight on principle. The result was global war.

But that war cannot be blamed on Neville Chamberlain. World War II, a germ in Adolf Hitler's furious mind, was coming no matter what Chamberlain did or failed to do, and if anything these negotiations serve only to underscore the enormous gulf between genuine level four/five hopes for decency and peace and the extreme, level three belief in predatory warfare that animated the Nazi regime during the Hitler years. The war in Europe was all Hitler's doing. While Chamberlain was certainly fooled, he at least went down a fool for a high and noble cause.

What Chamberlain's failure *does* exemplify for all future negotiators, however, is the absolute necessity of identifying the level — and thus the worldview — of all parties involved in any international negotiations. Level four/five, unmindful of the developmental facts, tends to believe that all people are alike in their honest desire for peace and tranquility — just like *them*, in other words. Yet history has time and again demonstrated the fact that this belief is *incorrect*. Many level three regimes do *not* want peace; they want *war*. And it was here that Chamberlain was hopelessly mistaken, while Winston Churchill, to his credit, accurately perceived Hitler for the predator he was.

In 1938, after a long series of negotiations between Hitler and Chamberlain, the French and British gave in to Hitler's next series of demands and allowed German troops to march into the Sudetenland, a portion of Czechoslovakia Hitler had long had his eye on. Chamberlain returned from the Munich negotiations waving the treaty in his hand, all the while promising the British people, "I believe it is peace for our time.... Peace with honor"— the profoundly mistaken words for which he will be forever remembered.[16]

Yet peace did remain a fleeting reality until, after signing a nonaggression pact with the Soviet Union which he would abide by only so long as it served his purposes, Hitler invaded Poland in 1939. By then Britain and France had had enough, and both declared war against Germany. Unfortunately, neither the French nor the British were prepared for the war they had just declared. The Germans, on the other hand — and in order to avoid the disastrous consequences of trench warfare which had brought them to their knees during the Great War — had devised an entirely new approach to the offensive.

Called blitzkrieg (lightning warfare), this new offensive called for the integration of infantry, tanks, artillery, and forward air support in a manner never before conceived. "Blitzkrieg was designed to so stun and disorient the enemy that it was unable to mount a response. Reliable two-way radio communication in the field and technologically advanced aircraft and armored units now made it possible to launch well-coordinated, fast-paced, mobile military operations that could quickly change course if needed."[17] If all went as planned, the German advance would never bog down or be forced back onto the defensive. Trench warfare, it was hoped, the principal reason for German failure in the First World War (at least, so it was thought in Berlin), would thus be avoided.

The idea of the blitzkrieg was to rapidly break through the enemy's weak points; cut off and encircle; fragment and destroy — in short, divide and conquer at high speed. Rapid advance and tight coordination meant everything to the effective implementation of the blitzkrieg. "Heavy aerial attacks focused on communications lines and transportation networks (airfields and grounded aircraft were the top priority so as to obtain immediate air superiority). Massed tanks simultaneously charged forward making deep incisions into enemy lines followed by reconnaissance, anti-tank, artillery, engineer, and signal units as well as armored personnel vehicles."[18]

In Poland, Belgium, and France blitzkrieg worked beyond even Hitler's most ambitious projections. The German advance through France would gain the English Channel in only one week, and on June 14, 1940, the Germans marched into Paris unopposed. It was an unprecedented victory, and a victory that only added fuel to the fire of Adolf Hitler's predatory fantasies.

Hitler ultimately dismissed the thought of invading Britain as operationally implausible and turned his attention instead toward Russia, the true goal of his predatory reverie. On June 22, 1941, Operation Barbarossa was launched, 3200 aircraft and 20 mechanized divisions totaling some 8,000 tanks crashing across the Russian border. The nonaggression pact between the two nations had been suddenly, violently, and unilaterally cancelled.[19]

Throughout that late spring and summer the German assaults sped through western Russia. The offensive, comprised of three separate mechanized columns, overwhelmed then devoured the spirited but meager Russian resistance and penetrated deep into Russian territory. The Russians, routed into disorganization, were no match for the invaders, and by late summer it appeared Operation Barbarossa was headed for complete success by early fall.

Overruling his high command, however, Hitler ordered the offensive to concentrate on Russian industry, for the moment forgoing the heart of the Russian regime, Moscow. But things did not go as planned. With October the rains came, turning roads to mud and bringing the German advance to a grinding halt. Suddenly time and weather began working against the Germans. "But as October turned to November, the steadily dropping temperatures brought misery to the lightly clad German troops and chaos to their supply system. Hitler had been so confident of achieving victory in 1941 that he had not provided for a winter campaign."[20] It would not be the first time overconfidence had been the undoing of an army.

Snow began to fall. Still, the Germans crept toward Moscow. The temperatures dipped steadily — ten, twenty, thirty below zero — and the German infantrymen, sporting only summer uniforms and equipment, began to freeze to death. By early December, however, they were finally within sight of Moscow.

As the Germans inched ever closer to the Soviet capital through the bitter cold of the Russian winter, on the other side of the globe a Japanese carrier group silently inched its way across the Pacific Ocean toward Pearl Harbor in the Hawaiian Islands. The Japanese had designs on most of the Pacific, and their only true adversary in that quadrant of the world was the United States. The American Pacific Fleet lay at anchor at Pearl Harbor, and the Japanese high command firmly believed that if they could deal a death blow to the American fleet they would have at least six uninterrupted months to run free across the Pacific, conquering as they pleased.[21] By then, they assumed, the Americans would have little choice but to accept the new order of things in the region and settle for peace.

The Third Reich and the Imperial Japanese Empire, two level three militarily oriented societies, were, in December 1941, both functioning in terms of predatory, mythic conceptualizations that were in large measure out of touch with military reality. Hitler, ignoring the reports from his general staff, insisted on pushing Operation Barbarossa beyond the physical and logistical limits of both his men and his equipment. The result would be disaster. But Adolf Hitler was by his very nature immoderate, thus when his plans went awry, they seemed not just to fail, but to implode.[22]

As far as the Japanese were concerned, the odds of their defeating the United States in an all-out war were at best ten to one. Yet military myths often run roughshod over truth, and in 1941 Tokyo, self-serving martial myths were on the gallop. The surprise Pearl Harbor strike was designed by the Japanese to eliminate the United States from the Pacific theater for at least six months, thus allowing the Japanese to secure a new, far greater empire. The surprise attack had the exact opposite effect. "Pyrrhus the Epirot would have understood Pearl Harbor. For he learned against the Romans that to win against such an opponent is to lose.... The surprise attack provoked a rage like no other in American history."[23] What the Japanese high command initially conceived as a complete and total success would, some four years later, require radical reconsideration, for by then two million Japanese men, women, and children would be dead and the Japanese Empire in ruins.[24]

As pointed out previously, level four democracies tend to be slow to anger, slower still to war; but once aroused they can be formidable opponents. The Japanese high command made the classic level three mistake of confusing the reluctance to fight with weakness and paid a terrible price for the error. "John Toland makes the point that the Japanese attack was rooted in the same impulse which brought forth their favorite literary form the haiku, a seventeen-syllable equivalent of a coup de grace. That was the problem. A sumo or kendo match might be finished by a sudden burst, but not the Americans. To them Pearl Harbor marked the beginning, not the end."[25]

Recall that, historically, level three empires have always tended to be racist if only in the sense that they naturally fancy themselves superior to any "obviously inferior" outsider who is to be killed, used, or enslaved as need would have it. But in World War II this impulse was magnified in both Germany and Japan to monstrous proportions. In Germany it was the inflammatory, racist rhetoric and policies of Hitler and his henchmen that drove the

Japanese attack at Pearl Harbor, Hawaii, December 7, 1941. While tactically successful, the surprise Japanese attack would prove to be a strategic blunder of the first magnitude and the death knell of the Japanese Empire (Library of Congress).

German nation into a genocidal frenzy the world has yet to fully digest — the Holocaust. But Japan was not far behind. The Japanese soldier, schooled to firmly believe that it was a dishonor to be taken alive, considered anyone who surrendered beneath contempt. Coupled with a natural sense of racial superiority, this attitude led to horrific atrocities all across Southeast Asia. While the Japanese had no official genocidal agenda to speak of, often their unit atrocities were so grotesque, systematic, and barbaric that they were often interpreted by those who stumbled upon the aftermath as being a natural expression of official policy.[26]

This is not to say that American, British, or other Allied troops were incapable of atrocious behavior; they were. While we may speak almost philosophically of war "between levels," for the combatants at the front war is always a fight to the death, a crazed, frantic ordeal of survival; and both physically and psychologically they respond accordingly. But individual, or even unit, atrocities cannot be compared to widespread, government-sanctioned — or, more precisely, *ordered*— slaughter of innocent captives or the wholesale warehousing and annihilation of an entire race of people.

Thirteen. The Great War, Act II

In World War II the world witnessed the wedding of a perverse, level three warrior mentality with the state-of-the-art level four technology of the day. The result was catastrophic. Stanton Coblentz sees it with remarkable clarity: "In the Second World War, however, we had the spirit of the Mongol tribesman, and in addition the miraculously effective new tools of annihilation — in other words, the mood and social morality of the barbarian governing the products of science."[27] Exactly. And that barbarian morality anchored in mythic notions of racial superiority would produce a trail of atrocity and genocide around the globe. Because of the wide-ranging atrocities committed by both the Germans and the Japanese — which generated a kill or be killed attitude in most Allied troops — World War II devolved into a particularly savage contest, and a contest that would ultimately claim over sixty million lives.[28]

Trench warfare, the plague of the Great War, would not be seen per se in World War II. In Europe, Hitler's blitzkrieg would sweep through France, Norway, and the Netherlands in weeks only, successfully eliminating the large-scale infantry confrontations that, in the Great War at least, required the combatants to ultimately dig for cover. Speed through mechanization was the key. After the failure of Operation Barbarossa, and later the D-Day invasion of Normandy by the Allies, the Germans would be fighting almost constantly in the retrograde and against overwhelming odds.

The Pacific Theater, on the other hand, would witness a continuation of the violent methods of trench warfare, but due to the island-hopping nature of the contest, on a much more limited scale. Elongated lines of trenches were simply impossible in the Pacific Ocean, yet on those islands where the Japanese and Americans met in combat, the fighting was some of the most savage in history. Dug in, the Japanese resisted almost to a man, and were slaughtered almost to a man. An American Marine, for instance, recalled the fighting on Saipan:

> A Jap jumped and grabbed one of our men. What the enemy would do is grab someone and hit his grenade on his own helmet [to trigger it] and then hold the grenade between them so they would both be killed. The marine grabbed the Jap, pinning both his arms so he could not start the fuse of the grenade. And they were dancing around, and somebody else went up and shot the Jap in the head with a forty-five. We called that boy "Dancer" from then on, because it had looked like they were dancing. He later got killed on Tinian.[29]

The initial Japanese expansion swept over the Philippine, Solomon, and Gilbert island chains, consuming in the process Singapore, New Guinea, Indonesia, Guam, and Wake Island. But after the American naval victory near Midway Island in June 1942, the soon-to-be short-lived Japanese Empire began to implode, island by island. Their carrier force decimated, the Japanese were forced to fight on the defensive and against overwhelming odds for the remainder of the war, a ready prescription for disaster.

As the American counteroffensive inched ever closer to mainland Japan, the Japanese response grew ever more fanatical, ultimately relying on the kamikaze, a suicidal aerial tactic that proved every bit as effective as it was desperate. The hope for the Japanese high command was that one well-armed plane might, with an enormous amount of either skill or luck, take out one American ship. One plane for one ship was in fact a highly efficient kill ratio if it could be made to work, and the initial onslaught of suicide bombers left the Americans reeling. Indeed, the kamikaze was a tactic that for a while proved highly effective and forced the approaching Americans into a mindset that justified the use of any ordnance, no matter how horrific, to end the slaughter.[30]

Meanwhile, back in Europe, Operation Barabarossa had bogged down as a result of the dreaded Russian winter and the vast Soviet landscape. Despite catastrophic losses in men and equipment, the Russians kept pouring in more of each against the hated Germans, whose reputation for butchery had turned the Eastern Front into a slaughterhouse. "This was to be a war without scruples, a predatory extravaganza."[31] Driven by Hitler's racist, genocidal agenda, the Germans led the way, butchering whole populations as they advanced deep into the heart of Russia. The Soviets naturally responded in kind, vowing to destroy the Germans or be destroyed themselves in the process. No accommodation was possible in such a situation. The war on the Eastern Front rapidly devolved into a fight to the death where absolutely no quarter would be given.[32]

In 1942 the Germans tried to regain the initiative by launching another offensive — Operation Blue — that would soon grind to a disastrous conclusion when in the following year the entire German Sixth Army was surrounded, trapped, and finally forced to surrender at Stalingrad. By 1943 severe logistical inadequacies were hampering German war efforts both in Europe and North Africa, and in 1944 the Allies finally launched the long awaited second front on the French coast at Normandy, forcing the German Wehrmacht into a retreat from which it would never recover. In Europe the war would end with Hitler hunkered down in his Berlin bunker, Allied forces closing in from all quarters. His thousand-year Reich had evaporated in just twelve years under a pile of debris and the sad reality of millions upon millions of human lives destroyed along the way.

The Pacific war would end in the blinding flash of the world's newest and most scientifically advanced weapon, the atomic bomb. "Kamikaze attacks and fanatical Japanese resistance on Iwo Jima and Okinawa," Robert O'Connell points out, "had convinced the president and his advisors, over the protests

Atomic bomb blast rising over Nagasaki, Japan. The atomic bombs dropped on Hiroshima and Nagasaki would end the slaughter of World War II but catapult the world into a postwar arms race of extraordinary proportions (Library of Congress).

of key scientists, that the bomb was the only alternative to an extraordinarily bloody invasion of the home islands."[33] On August 6, 1945, the first atomic bomb, named "Little Boy," was dropped on the city of Hiroshima, obliterating the target. A few days later a second bomb was dropped on the city of Nagasaki.

This atomic ordnance was not just another new weapon — far from it. In the giant mushroom cloud, the image that would soon become emblematic of nuclear horror, something was spotted that had never been seen before. It is said that the scientist Robert Oppenheimer, one of the bomb's chief architects, upon witnessing the bomb's initial test explosion in the New Mexico desert quoted a line from the ancient *Bhagavad Gita*: "Now I am become death, destroyer of worlds; waiting the hour that ripens to their doom."[34]

Not only were Oppenheimer's words in a sense visionary, they were literally on the mark. In the mushrooming nightmare of nuclear holocaust the original death terror, projected outward onto the world so many thousands of years before, had at long last come full circle, made manifest for all to see by the work of Einstein, Oppenheimer, and Urey. As O'Connell observes, "A nuclear explosion is impossible to ignore."[35] And, we might add, impossible not to grasp on some instinctual level. Nuclear warfare is not a strategic conception for a new form of international combat so much as a prescription for utter annihilation. In the frightening vision of a few hundred (or thousand) mushroom clouds rising high in the air there are no heroic deaths to be seen at all, only destruction and mass death — endless, universal.

Only seven years after the detonation at Hiroshima, the first hydrogen bomb would be tested in the Pacific Ocean, releasing a destructive force 1000 times greater than that of the original atomic weapon. In a very precise sense, nuclear weapons represented not another advancement in the evolution of weaponry but rather the "destroyer of worlds," as Oppenheimer had observed — and, as such, both the logical and psychological end point of twelve thousand years of human warfare. No longer could death be denied, projected, or extroverted. War no longer meant death for the other, but a mutually assured destruction that had become utterly apparent in the image of a thousand spiraling mushroom clouds that would consume not just the enemy but everything. In just over one hundred years level four science and technology had created an endgame of catastrophic proportions.

Homo sapiens sapiens had suddenly stumbled upon a strange, new, and frightening landscape. For humanity the choice was stark but simple: mature or perish.

Fourteen

Cold War Conflict

Almost immediately after the conclusion of World War II, the victorious Allies split into two adversarial camps that would, for at least the next fifty years, compete on a global basis. Essentially, this competition was between the Western democratic nations, led by the United States, and the Communist bloc, controlled by the Soviet Union, both sides employing any number of surrogate states along the edge of their mutual fault line as both buffers and antagonists. Under a daunting, and increasingly lethal, canopy of nuclear weapons, this competition would fortunately remain one of ideology and arsenals, and not an open and outright conflict in the classic sense. The "Cold War" had been ushered in.

The United States would hold a nuclear monopoly only until September 1949, when the Soviets conducted their own atomic test in the Pacific. From that moment forward both sides would engage in an arms race that would reach extraordinary dimensions. While the nuclear menace would ultimately prohibit each side from pulling the atomic trigger — although they would come dangerously close to it on more than one occasion — twelve thousand years of warfare had become so deeply ingrained in human thought and behavior that the urge toward war had to be addressed in some manner. So instead of war outright, new delivery systems and vast arsenals were accumulated on a competitive basis over the coming decades, arsenals so immense and hyperlethal that they simply defied all logic. "It was estimated that the United States and the Soviet Union between them had more than 50,000 nuclear warheads. Britain, France, and China together had 1,000 to 2,000. The destructive potential of these accumulated weapons was all but incalculable — certainly exceeding one million Hiroshimas."[1]

Oddly enough, this overabundance of weaponry did not increase the likelihood of war but actually served to decrease the possibility. With the extraordinary overabundance of weaponry available to both sides during the cold war, open warfare had suddenly been rendered obsolete. For to actually contemplate a nuclear exchange as a viable tactical strategy was to contemplate an insane proposition.² With the advent of nuclear ordnance, weaponry had not only outgrown its usefulness, it had outgrown war itself.

It is thought that ideology was the dividing element between these two hostile camps — communism vs. democratic, capitalistic systems — but that was true only in the most superficial sense. While democracy was a maturing idea driven by a thousand different minds (and continues to be), communist thought was the product of only a very few contributors, principally Karl Marx and, later, Nikolai Lenin. Both communism and democracy were thoroughly grounded in level four patterns of thought, democracy being the political expression of the ability to take the place of the other, communism correctly grasping the significant unfairness and inequities that existed, and had existed, in the distribution of wealth since there had been such a thing. Marx's *Communist Manifesto,* then, was a theoretical attempt to both make sense of and correct those inequities; and it was clearly communism's claim to a sort of utopian fairness that represented its principal appeal to the millions of poor and politically disenfranchised around the world.

The problem with communism was not Marx's attempt at equity, which was entirely commendable, but the fact that his theoretical view of the world was fundamentally flawed. As Ken Wilber points out, one of Marx's great mistakes was to take all the truths revealed in level four introspection and reduce them to "their lowest common denominator ... to material productions and material values, and material means, with all higher productions, especially spirituality, serving only as the opiate of the masses."³ Marx, like Hegel before him, understood that the world seemed to be developing in some natural way, but he profoundly misunderstood the nature of that development.

Marx had adopted Georg Hegel's historical dialectic, or the idea that history was unfolding in accordance with a natural, inevitable formula, the nexus of which for Marx became the working class, or proletariat. "Marx removed from Hegel's theory the assumption that nations are effective units of social history ... and replaced the struggle of nations with the struggle of social classes."⁴ Liberal revolutions, such as the one in France that had toppled the monarchy, represented to Marx simply the first phase of an ongoing struggle, the second, and last, phase of which he believed would produce a proletarian revolt that "would not merely transfer the power to exploit, but would abolish exploitation."⁵ The revolution of the proletariat, then, was to Marx's way of thinking the omega point and ideal state toward which history had been laboring for thousands of years. But this conception was radically flawed, as will soon be told.

Recall that, generally speaking, philosophy split thousands of years ago, one branch leaning toward human development as the key to ultimate human success, the other taking the romantic path and looking backward toward

Eden in the hope of somehow resurrecting that original human goodness that the forces of civilization had managed to crush. The historical dialectic was essentially a developmental conception, while an omega point of an ideal proletarian state which would arise once the capitalist economic system had either dissolved somehow of natural causes or had been destroyed outright, was in fact little more than a romantic fantasy. Thus Marx was in essence attempting to look both forward and backward at the same time, utilizing a developmental scheme to deliver a romantic end point, and this in a nutshell was Marxism's great romantic flaw. And as we saw with the French Revolution, when history will not cooperate with the romantic agenda, the romantic often finds it necessary to hurry history along. This generally begins with little more than intellectual bullying, but it often ends with murder, concentration camps, and ultimately genocide because simple bullying always fails to deliver the desired result.

Marx lived in a Europe that was experiencing the initial influences of level four growth but where social classes were still very much hardened structural realities. What he did not understand was that in the democratic nations of the future, social classes would become far more porous, while *individuals* would become remarkably fluid in relation to those classes. Thus the life trajectory from working class laborer to bourgeois capitalist would become not at all uncommon.

In that sense, the revolution of the proletariat that Marx envisioned was, in fact, nothing more than a philosophical fantasy, no more predicated upon scientific inquiry than is much theological speculation; and just as with theology, believers in Marx's revolution accepted its inevitability on the basis of faith, not reason. To the true believer then, communism was revealed truth, not truth reasoned, and the virtually evangelical nature of Marxism has not been lost on many of its critics. "Historical necessity, therefore means [to Marxists] not merely cause and effect, or desirability, or moral obligation, but all three at once—a kind of cosmic imperative which creates and guides human interest and human calculation and makes them its servants. But while Calvinists called this theology, Hegelians and Marxists call it science."[6]

Always anxious for inclusion in the modern academy as a genuine product of scientific thought, Marxism, much like modern Creation Science, was in reality nothing more than a myth masquerading as science. Science is based upon the scientific method, the rational examination, testing, and refinement of a hypothesis, and a process in which neither Marxism nor creation science, despite their claims, have ever appeared particularly interested.

Marxism, then, as rational, scientific, social, and economic theory, because of its gross reductionism, theoretical confusion, and lack of rational inquiry, was flawed beyond repair and intellectually dead on arrival. That did not stop

Nikolai Lenin, and later Mao Tse-Tung, however, from adopting large portions of Marx's theory, then attempting to kick, hammer, and glue it onto two of the most backward, agrarian societies in existence at the time — Russia and China. This grafting of flawed, level four philosophy onto a primitive, level three societal base would prove one of history's most curious intellectual and social disasters.

The first move was Lenin's, but as George Sabine points out, while Marx was utterly confident as to the direction of history, how exactly this revolution was to accomplish its goal of a proletarian utopia remained fundamentally a complete mystery: "Of positive or constructive ideas for this change it [the revolution] had remarkably few, for its energy had gone into making a revolution, not into making a program."[7]

Still, like a distant drumbeat, for Lenin's revolution there was the ever alluring call of the proletarian utopia. "It had indeed a goal: it would construct a collectivist economy and a socialist government.... It was the goal that formed its principal tie with Marxism, that remained the constant objective of its forced improvisations, and that required violent manipulation of the society in which it had to conduct its experiment."[8] And it can be understood that "forced improvisations" and "violent manipulation" of societies are in fact perverse, level three methodologies, and not the methods of level four or higher.

Joseph Stalin, a tyrant of the first order, would then turn that "experiment" into not a worker's utopia as Marx had predicted but one of the most predatory, level three empires in world history, brutally liquidating millions of Russian men, women, and children in the process. "The regime was socialist only in the sense that the nation owned the means of production; its realities were political absolutism and the imperatives of industrialization.... The consequence was that its policies had little perceptible relationship to the theories that it professed, which often seemed a mere facade for conventionally nationalist and imperialist behavior."[9]

Exactly. The Soviet Union was communist in name only. In almost all respects, its goals, methods, and worldview were typical of traditional level three predatory empires. Despite all the intellectual puff and systematic hostility between Russian communism and German Nazism before and during World War II, the truth of the matter is, the two regimes were driven by paranoid megalomaniacs who were far more alike in their characteristics, goals, and methodologies than they were dissimilar.

Much has been made of the collapse of world communism, but the historical reality is that communism in both the Soviet Union and China was never more than the thinnest veneer coating the everyday political dealings of otherwise traditional level three, predatory regimes. The Soviet Union's demise was hardly occasioned by the theoretical flaws inherent in Marxism, numerous though they were. The Soviet government collapsed because it was an incompetently managed, level three society hijacked by a clique of second-rate thugs and ideologues far more interested in the traditions of power than in any true Marxist goals. If we simply think back to our rough description of level three societies — imperial, culturally agrarian, religiously fundamental, and socially stagnant — it is easy to see that Stalin's Soviet Union had far more in common with Xerxes' Persia (minus the splendor and wealth, that is) than with any utopian vision Marx ever contemplated. The result was an eighty-year political, intellectual, social, economic, and ecological nightmare.

Today it seems that communism, once the darling of European intelligentsia, is championed by only the most naive and deluded academics, its clarion call of a worker's utopia through proletarian revolution forever discredited by a century of ruthless oppression, genocide, and failure. Today the old Soviet Union struggles mightily to grow beyond the disaster left behind by its communist rulers, a growth greatly affected by Russia's historic level three outlook and culture. Ken Wilber explains:

> Although ostensibly a modern state, its infrastructure was more substantially that of a blue [level 3] ancient nation, with totalitarian rule, one-party dominance, command economies, and collectivistic ideals. Because an orange-meme [level 4], individual-initiative-driven, capitalistic market cannot develop under those circumstances, when something resembling a market economy was abruptly introduced, the ancient nation did not evolve forward into an orange [level 4] modern state but in many ways regressed back to a red [level 3/2] feudal empire, rife with warring gangs, criminal warlords, and a Russian mafia that controls much of the market.[10]

Much of Marxism's initial insight was corrected in the Western democracies by a host of natural adaptations — public education, labor unions, civil rights laws, an enormous growth in colleges and universities, etc.— that softened the old classes themselves, while facilitating movement through those classes for any motivated individual. The result was that Marx's initial negative critique of capitalism, which was accurate so far as it went, has been greatly rectified through natural market and social adjustments.

With the collapse of communism across the globe, the world was left with essentially one superpower — the United States — and a radical imbalance of military and economic might. American nuclear and technological arms superiority has made war against this military giant, at least on a traditional transnational basis, almost unthinkable.

As humanity enters the twenty-first century, traditional, predatory level three war such as Iraq's most recent invasion of Kuwait, seems at long last to have fallen on hard times. The leading nations of the world, through the agency of the United Nations, seem to have, at least tacitly, accepted the proposition that aggressive, territorial acquisition through the use of arms is now a thing of the past. If true, this would represent a huge step forward in the eradication of predatory warfare and should be loudly applauded by all those interested in a stable, peaceful future.

While the need, therefore, for traditional, large-scale military operations seems to be scaling down, the world left behind by the communist collapse is one in many respects far from ideal for the rapid transition to democracy. As pointed out previously, today some 70 percent of the world's population remains at or below level three in terms of development, a fact that must be plugged into any formula for world advancement, prosperity, and the natural eradication of warfare. Wars continue to be fought, of course, but they tend to be more contained and less combustive than in previous decades.

Likewise, while arsenals grew geometrically during the post–World War II period, the implements of war, despite considerable augmentation, have recently advanced very little. While it seemed to many that the race toward war was accelerating at an alarming rate, the pace of technological advance has actually slowed down. As pointed out by Robert O'Connell, the truth of the matter is, the vast majority of weapons and systems in today's arsenals were already on the table by the end of World War II. The ballistic missile was crafted by German engineers, and by war's end they had designs in the planning stage for models of intercontinental range. Also on the drawing board were missiles that could be launched from systems towed by submarines. The first German jet fighters were already in the sky as the war swept forward into its closing stages, and prototypes with both swept and delta wings had already been engineered. Indeed the first, primitive prototypes for the modern cruise missile had been developed in Germany before the war had ended.[11]

For the most part, the armament advances during the second half of the twentieth century were refinements, most notably in target acquisition of already existing systems, and not the groundbreaking work that ushers in entirely new schemes of war making. In that sense it seems the world today is on the cusp of a giant leap forward into level four/five education, wealth, democracy, and prosperity, or a gigantic stumble backward into level three tyranny — both religious and secular, or worse. While conflicts still occasionally ignite along the ragged seam between these two levels like sparks along the edge of a grinding wheel — most notably in the Middle East — so far they have remained nonnuclear in their violence and local in their scope. So there is at least some rational hope that warfare, as the world has known it for almost twelve thousand years, may, with the continued development of level three societies toward level four democracy, become a thing of the past.

In just over fifty years the prospect of a massive nuclear conflagration appears to have been, if not nullified entirely, at least greatly diminished. That is no small accomplishment. Yet as humanity continues to develop, war alters to keep pace; that has been war's great constant since its inception. Just what the face of warfare might look like as *Homo sapiens sapiens* continues to develop in the coming decades is the topic of the next chapter.

Fifteen

The Evolution of War

The thrust of this inquiry has been that war is essentially a product of human psychology, and that human psychology has been continually developing over the span of the last 12,000 years. Not only, therefore, has the character of war — the manner in which wars are fought — been altered in accordance with this ongoing development, but the very fundamental motivation to go to war to begin with has also evolved as the general level of psychological development has changed. If this is true, and to further demonstrate the validity of these claims, we should not only be able to identify these shifts in the methods and motivations of warfare throughout the historical record, but also to make reasonable predictions predicated upon the characteristics of the more advanced developmental stages — levels that have been systematically identified but have yet to appear to any large degree in the general population. And, as we shall see as our inquiry continues, that is exactly the case.

It may be recalled that both levels one and two were incapable of war. This was not because these peoples were fundamentally peaceful; it was entirely a function of their *lack* of development. Sparrows don't go to war either, but we don't consider them peaceful in any meaningful sense. These first two levels existed during a time before cities, before what we today call civilization. Ken Wilber characterizes level two in this manner:

> Thinking is animistic; magical spirits, good and bad, swarm the earth leaving blessings, curses, and spells which determine events. Forms into *ethnic tribes*. The spirits exist in ancestors and bond the tribe. Kinship and lineage establish political links. Sounds "holistic" but is actually atomistic; "There is a name for each bend in the river but no name for the river."[1]

This is not to belittle these early humans. The human experience had to start somewhere, and in many ways the challenges these people faced and overcame make our own today seem minuscule. But they did not war, and they did not war because they had not yet developed the expanded worldview and understanding of themselves that would ultimately trigger the march to conflict.

That expanded capacity would appear with the dawn of level three around the year 10,000 B.C. Humans would develop a greater sense of themselves, which would naturally be accompanied by a greater comprehension of death. As we have seen, this death fear would then be violently projected by the new, fragile self, and war on a wide scale would for the first time appear. For simplicity's sake I have called this stage level three, a phase Wilber now actually divides into two separate stages.

So I will present Wilber's characterizations simply as early stage three, then as latter stage three. The early stage would appear roughly around the year 10,000 B.C. and dominate world events until approximately A.D. 1200 This does not mean, of course, that all individuals during this time period aspired to level three only and no further. Indeed, many remained below level three, while any number of people moved beyond it. Level three is simply the rough average for that given period of time, and this applies to all time periods we have discussed in this inquiry. Note how closely the wars fought during this early period reflect the general psychology of the era. Here now is Wilber's characterization of early level three:

> First emergence of a self distinct from the tribe; powerful, impulsive, egocentric, heroic. Magical-mythic spirits, dragons, beasts, and powerful people. Archetypal gods and goddesses, powerful beings, forces to be reckoned with, both good and bad. Feudal lords protect underlings in exchange for obedience and labor. The basis of *feudal empires* — power and glory. The world is a jungle full of threats and predators. Conquers, outfoxes, and dominates; enjoys self to the fullest without regret or remorse; be here now.[2]

This, clearly, is the world of Alexander, Caesar, Hannibal, all the great Khans, Attila the Hun, and many others. Notice how, without even trying, Wilber provides almost the entire vocabulary for predatory warfare: powerful, jungle, gods, obedience, predators, dominates, impulsive, feudal, glory, and, of course, heroic. This is the period in which war had its birth, and in which its initial patterns of behavior were laid down. Here the warrior ethos would emerge, the Homeric codes formulated and given life — it seems — everlasting.

While for us modern people this seems like ancient material, nothing could be further from the truth. Even today in the United States, for instance, the Marine Corps uses a recruitment ad on television that features a young, heroic Marine want-to-be climbing a mythic mountain to do battle with an archetypal, fire-breathing dragon. This is a commercial cut straight from the cloth of early level three symbolism; an ad that is effective because it appeals directly to the original warrior ethos, and an ad Alexander, were he alive today, would have no trouble appreciating. We are not so advanced as we like to think. Our world is still rife with early level three thinking, symbols, and behavior. Where? Ken Wilber's answer follows:

The "terrible twos," rebellious youth, frontier mentalities, feudal kingdoms, epic heroes, James Bond villains, gang leaders, soldiers of fortune, New-Age narcissism, wild rock stars, Attila the Hun, *Lord of the Flies*. 20 percent of the population, 5 percent of the power.[3]

Early level three would then slowly develop over time a more mature, more elaborate comprehension of life, and warfare would evolve accordingly. During this period, which continues throughout much of the world to this day, the old feudal kings would morph into more stable, imperial monarchies (or tyrannies), but these new forms of government would remain profoundly predatory. What was the psychology of this late level three period like? Ken Wilber has this to say about it:

> Life has meaning, direction, and purpose, with outcomes determined by an all-powerful Other or Order. This righteous Order enforces a code of conduct based on absolutist and unvarying principles of "right" and "wrong...." Following the code yields rewards for the faithful. Basis of *ancient nations*. Rigid social hierarchies; paternalistic; one right way and only one right way to think about everything. Law and order; impulsivity controlled through guilt; concrete-literal and fundamentalist belief; obedience to the rule of Order; strongly conventional and conformist. 40 percent of the population, 30 percent of the power.[4]

Here is Germany, its troops marching off in lockstep to happily initiate two world wars bound, it was thought, to conquer Europe while proving their own (Germanic) cultural and racial superiority. Here are both Stalin and Hitler, "liquidating" millions of human beings who were considered subhuman, defective, or simply at odds in some ill-defined manner with the new Nazi or communist Order. Here too is the Japanese Imperial Army, butchering Koreans, Chinese, and much of Southeast Asia in accordance with their own level three notions of racial superiority.

But the list hardly ends there. The Inquisition, the Red Guards, the Samurai ethos, Confucian China, Manifest Destiny, English Royalty, Islamic Fundamentalists, and Puritan New England all serve as boilerplate examples of this level in terms of thought, psychology, and warfare. God is often invoked at this level to quash any difference of opinion or to condone any exercise of the state, a state which is often confused or fused with God in the first place. As late as the 1940s, for instance, the Japanese emperor was considered by the people of Japan to be literally divine. Wilber says, "This is why most religions, centered on the blue meme [late level 3] have *caused* wars, not prevented them. Not only have religions caused more wars than any other force in history, they did so in the name of an intense love of God and country. Their love was ethnocentric, dispensed freely to true believers and the chosen people, and death to all others in the name of that love and compassion."[5]

This late level three mentality would rule the world until the emergence of a level four outlook, which appeared initially with the Renaissance, later blossomed more fully in the Enlightenment, and finally burst forth in the American Revolution. But level four would have to extricate itself from the clutches of level three groupthink, a process that would initiate revolutionary wars predicated no longer upon the predatory acquisition of land or riches but rather upon the acquisition of freedom, equal rights, and, in some cases, new national identities, Italy and Germany serving as examples of the latter. As previously pointed out, level four, due to its democratic institutions, tends to shy away from predatory warfare. Rationality, industry, science, technology, and democracy are the hallmarks of level four. Where can we spot level four development?

> The Enlightenment, Ayn Rand's *Atlas Shrugged*, Wall Street, emerging middle classes around the world, cosmetics industry, trophy hunting, colonialism, the Cold War, fashion industry, materialism, secular humanism, liberal self-interest. 30 percent of the population, 50 percent of the power.[6]

Level 3 would first initiate war, and level four would then change war's central motivation from one of death-dealing and acquisition to the securing of freedom, national identity, and civil rights — an enormous shift. Many of the conflicts that have raged over the last 20 years can be seen as clashes between levels three and four, e.g., Bosnia, the Israeli-Palestinian conflict, 9/11, and both Iraqi wars. But, interestingly enough, the evolution of war does not end there.

Today, at least in the Western democracies, the beginnings of a fifth stage have been identified, and this stage has also had an effect on how nations see and utilize their military resources. This stage goes beyond level four's capacity to simply take the place of the other. Ken Wilber characterizes stage five as follows:

> Communitarian, human bonding, ecological sensitivity, networking. The human spirit must be freed from greed, dogma, and divisiveness; feelings and caring supersede cold rationality; cherishing the earth, Gaia, life. Against hierarchy; establishes lateral bonding and linking. Permeable self, relational self, group intermeshing. Emphasis on dialogue, relationships.... Subjective, non-linear thinking; shows a greater degree of affective warmth, sensitivity, and caring, for earth and all its inhabitants.[7]

Level five naturally has enormous potential, of course. But so far its arrival has been greeted with decidedly mixed results. Because of this stage's "affective warmth, sensitivity, and caring, for earth and all its inhabitants," for instance, level five tends almost always to be deeply antiwar and profoundly antimilitary. This stage tends, therefore — despite its professed caring for the earth and *all* its inhabitants — to be deeply contemptuous of anything arising from levels three and four, the levels it seems to hold responsible for almost every evil the world has ever seen. Level five views levels three and four as being cold, repressive, insensitive, militaristic, brutally rational, and dead wrong on almost everything. Sadly, level five does not appear to get along with level four any better than level three does. Level five as well, because of its enhanced sensitivity, tends to be easily offended, especially by anything that smacks of excellence, rankings, war or conflict.

Examples of this hypersensitivity abound. "Live Free or Die," for instance, has been the official motto of the state of New Hampshire since 1809, but this relatively harmless saying rankles level five proponents to the point that they have initiated numerous attempts to have it removed from the state flag. The saying is, of course, revolutionary level four in character, a defiant declaration of independence; but it is the bellicose insistence and militaristic overtones of the saying that rankles level five virtually no end. Thus the ongoing feud. Another example of this conflict is the almost endless antipathy of college and university faculties (comprised primarily of level five professors and teachers) to the mere presence of ROTC cadres (levels three and four) on their campuses. In all fairness, it should be recalled that, just as both the level three self and the level four ego emerged in fragile, immature form, level five is emerging initially in its juvenile configuration as well, a configuration given at times to somewhat petty and childish displays of temper and ego.

In the field of education, for instance, this immaturity has proven particularly discouraging. As Ken Wilber points out, "Middle and higher education in this country [the United States] now actually *encourages* ethnocentric identity politics, gender essentialism, racial identity, and the politics of self-pity — all part of 'rich diversity.' History is being taught as self-esteem therapy: not what happened where and when, but what immoral slugs they all were compared to you. Using the values of the liberal Enlightenment, you condemn all previous history, including the liberal Enlightenment."[8]

In other words, level five is actually encouraging a *regression* in behavior and culture, not an advance, as its proponents enthusiastically proclaim. Ethnocentric and racial identity policies and politics are the policies and politics of level three, attitudes that are finally discarded at level four with the emerging ability to take the place of the other. This regression is taking place because most level five proponents have no grasp whatsoever of human development and thus have actually confused regression with growth, simply because these regressive attitudes represent a move to something *other* than level four. So in a sense, with the emergence of level five we've seen one step forward, but two steps back.

What should occur at level five is an even deeper appreciation of human commonalities than usually present at level four. What we have gotten so far is an incoherent, "multi-cultural" Tower of Babel, foolishly employed to *disunite* what in some cases has taken hundreds of years to bring together. Because of level five's uncompromisingly antagonistic attitudes toward levels three and four, it has so far proven an impediment rather than a facilitator of positive world transformation. In that sense, this level is not currently producing the conditions that might eradicate war but the conditions that actually generate war in the first place: ethnocentric attitudes and identities. Where might we encounter elements of this emerging, level five thought?

> Deep ecology, postmodernism, Netherlands idealism ... Canadian health care, humanistic psychology, liberation theology, World Council of Churches, Greenpeace, animal rights, ecofeminism ... politically correct, diversity movements, human rights issues, ecopsychology. 10 percent of the population, 15 percent of the power.[9]

As pointed out, in many ways the emergence of level five has been a clumsy disappointment; but to be fair, in other ways it has had a profound and beneficial effect. The sensitive outreach inherent in level five, when not obsessed with symbolic trivialities, has already stretched military operations into configurations never before conceived. It is not coincidental, I think, that the very first peacekeeping and humanitarian forces have been mustered to the task during this era. Missions to feed and protect compromised populations, to restore order, and maintain the peace have, for the first time in history, been launched across the globe. While some of these missions have come up short due to their overly idealistic goals, many others have proven successful.

This is no minor alteration in the application of military force. It represents a significant shift in motivation, yet virtually no one seems to have grasped the fact that something significant has occurred. For the very first time in world history armed forces are being utilized not to conquer foreign lands or overthrow tyrannical oppressors — levels three and four motivations — but to care for and protect populations at risk. Clearly, this appears to be a motivation spurred by level five development, and a welcome one at that. As more of the world's population attains level five, it is entirely reasonable to assume that such missions will seem not only sensible but also absolutely critical to a world population growing closer with every passing day.

The developmental arc moves from an ethnocentric worldview to what Ken Wilber calls a worldcentric point of view: that is, from an appreciation of people and culture that is very narrow and limited to an appreciation that encompasses the entire world and all its peoples. This worldcentric perspective is fully able to appreciate the multitude of cultural differences that exist across the globe while never losing sight of the basic commonalities that unite all people, no matter their race, creed, color, or any other characteristic.

In that sense, if we simply follow this developmental dynamic to its logical conclusion, more and more people will be maturing to levels from which war will seem, except for purposes of self-defense, an illegitimate expression of national will. Just as slavery — an institution that existed happily all across the globe for thousands of years — was erased from the world scene in only a few hundred years with the emergence of level four morality, war may well go the way of the horse and buggy in a comparable period of time. This will happen, not because war will be outlawed or shouted down or kept at arm's length by a tenuous balance of power or a shaky network of international alliances but because humanity will simply outgrow it. And that future may not be so far off as previously thought.

Our inquiry continues.

Sixteen

The Future's Promise

The end of war will not come with pious admonitions to refrain from hostilities, legal machinations designed to prevent conflict, idealistic social policies intended to stamp out war's "root causes," or Draconian bouts of consciousness-raising aimed at the morally "inferior" many by the morally "superior" few, all historically well intended but often useless or even counterproductive avenues of approach. The elimination of war will come as a result of human development across the globe, and that development will evolve naturally once the proper conditions are in place. That is simply a fact, and that fact is simultaneously both the good and the bad news.

It is the bad news because, as previously discussed, almost 70 percent of the world's population still hovers at or below level three, experiencing lives that are ethnocentric by nature, and subsequently perfectly capable of supporting predatory warfare. Yet it is simultaneously the good news, because for many of these populations the leap outward is not such a great one, and in many cases that leap is already well underway.

It might seem logical that in order for war to subside the general average for the world's population would have to shift to level five or even a more lofty level — since level five itself, as we have seen, is having an awkward time of it, but that is not so, and that, too, is part of the good news. For while level five does offer "a greater degree of affective warmth, sensitivity, and caring, for the earth and all its inhabitants,"[1] it is not necessary to achieve that level of growth for war-making to be greatly curtailed. It will not require an enormous shift in psychological level for the world's population to eliminate war, just a modest shift of the general average from level three to level four. Why? Well, it can be recalled that it is at level four where the urge to predatory warfare begins to subside. The principle reason for this is that level four allows the individual to take the place of the other, and this simple but profound shift in perceptions rapidly ushers in a whole new worldview. Suddenly people across the globe, despite all our multicultural differences, are understood to struggle, suffer, and experience the world essentially the same way as we do, and from this basic understanding eventually flows a sense of equality, community, and expanded compassion. I can appreciate that person whom I have never seen or met before in some far-off land just as I can appreciate my son, friend, or next door neighbor simply because she is a human being.

Level four inaugurates a whole new way of thinking. A rational examination of ourselves and our world at this new level seems suddenly essential. Virtually overnight, then, the old level three systems of governance and belief are grasped as entirely inadequate. If, after all, we all deserve equal treatment, then a new form of government seems necessary, a government of laws not driven by the whims of men or women, and the infant democracy is born. The old warrior king is replaced by a new executive, an official no longer free to indulge his predatory instincts, but now bound and limited by *law*. The ultimate power of the state is no longer located in the unencumbered hands of the king or tyrannical ruler, but is now situated in the hands of the people. And the people are not, by and large, anxious to see their sons and daughters devoured in foolish wars.

Yet the people at this level also intuitively grasp the profound benefits of their new freedoms and governmental systems, and they will often unite in a ferocious defense of them if threatened. Thus level four societies appear largely to be slow to war but formidable opponents once aroused. As has been documented in previous chapters, it is not that democracies are incapable of predatory warfare. History demonstrates, rather, that democracies are simply far less *likely* to engage in predatory expeditions, and thus it is only logical that the more the world's population evolves the less likely predatory warfare will become. More important, concomitant with that evolution, the more vigorously will war be opposed by an expanding community of like-minded nations. It is to be hoped those two components will form the solid, level four basis for the ultimate extinction of predatory warfare.

It would seem, therefore, that to eliminate predatory warfare human society needs simply to implement the conditions that will rapidly induce level four development. That is true, but only partially so. What actually is needed is the implementation of conditions that will rapidly facilitate human development throughout *all* levels, not merely the conditions that favor one level or another. As Ken Wilber points out, "The prime directive of a genuine integral politics would be, not to try to get everybody to a particular level of consciousness (integral, pluralistic, liberal, or whatever), but to *ensure the health of the entire spiral of development* at all of its levels and waves."[2]

As the United States and the Western democracies enter the twenty-first century, for instance, there are essentially three levels functioning in most of them throughout their social fabric, and these three continue to function often as mutual antagonists. Recall from the previous chapter, as an example, that the recent emergence of level five has been flawed, and that this imperfection, because it is at an advanced level, threatens to stall, or at least seriously hinder, positive growth in general. Painted in very broad strokes, the three levels at work are three, four and five, and their unending feuds and squabbles are generally known, in the United States, at least, as culture wars.

The elimination of predatory warfare, as outlined herein, will require the development of the world's population from an average of roughly level three to an average of roughly level four; but if that growth is blocked or stunted, then war will continue unabated. So the problems associated with this imperfect transition must first be identified, then rectified.

The first problem with these culture "wars" is that all of the levels have something positive to offer, but none of them have the entire picture because none grasp the entire developmental spiral. So they are left, so to speak, hawking their own goods exclusively. Level three, as one example, has much positive to offer about family and spirituality, but generally only its particular sort of families and only its own version of spirituality. Level three, by definition, tends to be ethnocentric in its outlook and concrete operational in its mindset; thus its insights, while often correct as far as they go, tend also to be quite narrow.

Level five, on the other hand — and for reasons we will soon touch upon — has these days almost nothing constructive to offer about either family or spirituality. But level five is generally very good when it comes to genuinely reaching out to other communities and other cultures, something at which level three consistently exhibits incompetence. Each level has its own turf, so to speak, and, naturally enough, those turfs are defended vigorously. In that sense, then, the culture wars are in reality little more than turf wars, and all levels react violently whenever their turf is threatened.

Ken Wilber explains:

> [Each level therefore] reacts negatively if challenged; it lashes out, using its own tools, whenever it is threatened. Blue [late level 3] is very uncomfortable with both red [early level 3] impulsiveness and orange [level 4] individualism. Orange [level 4] achievement thinks blue order [late level 3] is for suckers and green bonding [level 5] is weak and woo-woo. Green [level 5] egalitarianism cannot easily abide excellence and value rankings, big pictures, or anything that appears authoritarian, and thus it reacts strongly to blue [late level 3], orange [level 4], and anything post-green [beyond level 5].[3]

Let's start with level five. Level five had its birth essentially with the Age of Enlightenment, that initial burst of level four insight which later evolved to level five. It was during the initial burst of level four insight that most of the basic rights of citizens were identified, and those rights would be expanded eventually to include other groups as level four itself matured and became less ethnocentric. Based on rational thought patterns, modern science emerged during this period and created its own methodology: the scrutiny of material phenomena through rigorous, objective testing, or what we know today as the scientific method.

Valid as is the scientific method, the investigation of material phenomena through the application of rigorous, objective testing by definition left the interior domain of the human experience out of the scientific equation. Notions such as free will, meaning, spirituality, compassion, values, character, intention, etc., suddenly had no valid scientific standing, not because they were invalid phenomena, but simply because they were not *material* phenomena that could be readily tested in the laboratory. X-rays might, for instance, locate a tumor in my chest, but they reveal nothing of my character. Thus notions like character, free will, and even spirituality became concepts level five over time ceased to truly believe in simply because they had become defined as unscientific, and as such they had the archaic, superstitious ring of level three about them.

So increasingly, as scientific materialism came to dominate the intellectual landscape, the interior domain was at first neglected, later disavowed, then finally forgotten altogether. Level four/five thought, that initial burst of what historically has been called the liberal worldview and emerging simultaneously with science during the Enlightenment, adopted this same material worldview, a worldview that worked well in the laboratory for a material science, but that would prove sadly deficient when it came to a wider grasp of reality:

> And liberalism, arising directly in the midst of this scientific materialism, swallowed its worldview hook, line, and sinker.... The only thing that is ultimately real is the ... material, sensorimotor world; the mind itself is just a tabula rasa, a blank slate that is filled with representations...; if the subjective realm is ill, it is because objective social institutions are ill; the best way to free men and women is therefore to offer them material-economic freedom; thus scientific materialism and economic equality are the major routes of ending human suffering.[4]

"The typical liberal, recall, does not believe in interior causation," Wilber points out, "or sometimes even in interiors for that matter."[5] This is the reason that to this day level five solutions for social problems are always material plans or material adjustments (welfare, food stamps, public housing, adjustments to the tax code, etc.). This is also why you will almost *never* hear the suggestion that someone ought to "pull themselves up by their boot straps" rising from a chorus of level five mouths, for such a thought is utterly foreign to a worldview that believes almost exclusively in external causation. Having no grasp of the interior domain, level five is left to craft solutions from one-half of the human equation only, and this is why these programs often wind up being well-intentioned but short-sighted flops. Millions and millions of dollars are spent, decades drag by, but little is gained, and in some cases, the original problem is even exacerbated. It is not that level five does not care. Indeed, level five is *the caring level*, and while its care and concern are heartfelt, its solutions are often critically flawed. If your heart is broken and you are on the verge of a mental breakdown, as a theoretical example, level five will buy you a new set of clothes and be sure you have enough money for the bus. "The typical, well-meaning liberal [level 5] approach to solving social tensions," says Wilber, "is to treat every value as equal, and then try to force a leveling or redistribution of resources (money, rights, goods, land) while leaving the values untouched."[6] Because of this, the emergence of level five across the United States and much of Europe has been flawed, flawed because this more developed level works exclusively within only one-half of the human equation — exterior causation.

Level three, on the other hand, appears to be a reasonably healthy expression of a much lower level, perfectly valid as far as it goes but it does not go very far. The principle difference between level five and level three is that level three understands the interior domain, but it generally fails to grasp the validity of the exterior, dynamic causes that can also affect the individual, something that becomes obvious with the more expanded level four/five outlook. Level three, therefore, tends to locate failure almost exclusively in the internal domain — an individual's character, pluck, or work ethic, as examples — while generally ignoring exterior causation. Level three's answers for most problems are therefore fished out of an ethnocentric pool of limited choices. If your heart is broken and you are on the verge of a mental breakdown, level three will tell you to pull yourself together, then suggest a trip to your church, synagogue, temple, or mosque. To solve a social problem, Wilber explains, "The typical conservative [level 3] recommends that we instill family values, demand that individuals assume more responsibility for themselves, tighten up slack moral standards (often by embracing traditional religious values), encourage a work ethic, reward achievement, and so on."[7] None of these solutions are necessarily wrong, by the way. They are, however — just as with level five solutions — only one-half of the puzzle.

Today, because it has virtually no concept of the human interior, level five calls many of the solutions favored by level three mentioned above "blaming the victim" and fancies any suggestion of their usefulness a barometer of intellectual inferiority and moral degeneration. Level five does this because its entire epistemology neglects the interior human dynamic and is predicated instead almost exclusively upon exterior causation. Solutions must therefore be provided (since individuals have no capacity to solve their own problems), generally by government, and demands never made upon the recipients, which would smack of foisting values on the unsuspecting victim.

When things go haywire within society, level five looks to exterior causation for the source of the problem, and finds them usually in global forces such as sexism, racism, capitalism, militarism, and other forms of social coercion. To the level five mind human beings are, like leaves tossed in a mighty wind, little more than creations of these global forces, never the creators of those forces themselves except, of course, for certain classes of human beings who have been branded the mandatory oppressors required by this model — Wall Street millionaires, as an example. Thus level five is forced by its own limited conceptions to conceive of human beings essentially as *victims,* and the wider net it casts the more victims it finds. Now, this is not to suggest that these exterior forces level five detects are not real. They are real. It's just that the level five world accounts for these forces and these forces *alone.* Since it can see and account for nothing beyond external causation, it cannot accurately address problems. Thus level five half solves everything, and half solutions generally turn out to be no solution at all.

Because level five refuses to see but half the problem, its worldview remains deeply flawed. Rather than address the flaw and work through the problem, level five has instead, like a con artist covering his tracks, invoked a sort of prohibition against any attitude, argument, or fact with which it finds itself uncomfortable, not at all unlike the censorship and book burning historically invoked by level three in order to avoid unpleasant facts. And because level five ignores half of the truth, there are many facts and assertions that today make it profoundly uncomfortable. This unfortunate stifling of legitimate dissent represents a regrettable regression.

While levels four and five initially emerged during the Enlightenment, a time when thinkers demanded evidence to support every idea, today level five has replaced the penetrating light of reason with a sad collection of orthodoxies, any deviation from which it finds intolerable. Indeed, "political correctness" is nothing if not a sad attempt to prevent any criticism of level five's blatant flaws, to gain through intimidation what it cannot hope to gain through genuine intellectual discourse; and political correctness — indeed, fear of the truth — is without question the saddest byproduct of this level's emergence to date. Political correctness is not the enforcement of some kinder, gentler world, as level five loves to tell itself, but rather an anti-intellectual, anti-democratic attempt to stifle free thought and to force through intimidation an agenda it cannot achieve through free, open discourse. Elements of level five have, as a result, literally turned many gains of the Enlightenment on their heads.

Level three, on the other hand, has no difficulty identifying problems within the interior domain, but is generally blind to even the most compelling social forces. Beyond those law-abiding citizens assailed by criminals (or by level five intellectuals) it rarely if ever considers anyone a victim, although we are all under assault by countless external forces daily. This lower level considers failure a function of elements like stupidity, or laziness, or defective morals only, and gives short shrift to the notion of any sinister global forces at work. Thus this less developed level tends also to half solve problems and, as demonstrated, half solutions are really no solutions at all.

The remedy to all this confusion is actually quite simple: We need to grasp the idea that human beings are all the products of both interior and exterior causation and that social solutions must take both into account in order to be successful. In short, I am the product of my will power *and* the environment in which I find myself, not simply one or the other. With that accomplished, level three will begin to grasp the greater picture while level five will truly become a magnet for all those levels below, which up until now have conceived level five as little more than an incomprehensible philosophy of victimhood. The developmental logjam at level five will be eliminated. Politics in the United States will take a leap forward, re-setting the intellectual bar at, say, level six; social and political intercourse will become, if not cordial, at least comprehensible; and social policy will be open to crafting solutions that actually succeed. If my heart is broken and I am on the verge of a mental breakdown, I may well need both a new set of clothes along with bus fare *and* some stern advice along with a trip to my spiritual counselor, not just one or the other.

Fortunately, many, many individuals have already managed this simple adjustment all on their own with no need for special psychological input or literary counseling, such as that just offered in this chapter. The reason for this is that the incorporation of both exterior and interior causation into our worldview is actually the most sensible, practical approach because it is one of the most basic realities of the human condition (most reasonably mature people automatically think this way). Unfortunately, these individuals often find themselves living rational lives in an irrational world, generally caught between two unreasonable poles of thought that today, in the United States at least, dominate social and political discourse. (This is not a book about Western politics, but it does need to be pointed out here that many individuals who refer to themselves as either liberal or conservative are actually working at much higher levels. Those who refer to themselves as conservative generally do so as a reaction to the foolish excesses of level five, while those who refer to themselves as liberal do so as a reaction to the foolish excesses of level three.) If this group of sensible, practical thinkers continues to grow, higher-level thought patterns might one day break through, and social and political discourse take a decided turn for the better.

What does all this have to do with war? It has everything to do with war. A more comprehensive understanding of human development is absolutely essential, not only for maximum growth but also in order to decrease tensions and hostilities across the globe. This grasp will allow for the creation of the necessary *conditions* that will lead to optimal growth at all levels. This optimal growth will then pave the way for a substantial developmental shift around the world, and that shift will usher in a proportional reduction in the probability of predatory warfare. It's that simple. That's how war will finally be eradicated.

Seventeen

Understanding Weapons

No inquiry into human warfare can be considered complete without a sensible examination of weapons; what they are, and what they have come to mean. This may seem at first a dry and rather technical topic, but in reality it is both a philosophical quest and, to a degree at least, a troubling excursion. This is because weapons predate not only human warfare but also humans themselves, that long line of protohumans that stretches back millions of years, and indeed beyond even that to the many primitive life forms that preceded them. In that sense, weapons have been an integral aspect of life's evolutionary journey and not simply humankind's adventures in self-destruction.

The dictionary defines a weapon simply as "any instrument or device used for attack or defense in a fight or in combat,"[1] a definition that is remarkably accurate but at the same time seems to rob the weapon of its depth of being and psychological authority. Weapons are indeed technical instruments, but they appear to be far older and far more integral to life's unfolding than simple tools. Robert O'Connell, who studied the relationship of weapons and the men who use them surely as much as anyone, offers this insight: "By virtue of an odd sort of consensus of opposites, weapons are generally perceived as matter-of-fact objects, mechanisms with little more symbolic and cultural significance than a pair of pliers or a stillson wrench."[2] This approach would be acceptable, O'Connell observes, if it did not fail entirely to address key questions regarding armaments and the seeming hold they have had on humanity across the ages.

The reason this deeper, or greater, understanding has not been achieved is that modern studies—social, psychological, even military—tend to view weapons simply as "things" or "systems" of things that have no meaning beyond their prescribed killing capacity. From this perspective, weapons themselves take on a distant, menacing aspect that seems almost alien to human culture, thus not making humans but weapons themselves the crux of the problem: get rid of those terrible weapons, and war would disappear. But weapons are simply a visible, tangible aspect of humankind's evolutionary trek, the tip of the iceberg, so to speak; thus, removing the weapons alone will accomplish nothing of substance while fooling us into a false sense of security. Weapons are in themselves articles of great significance, if only because they point beyond themselves to the very heart of creation, and ultimately to the heart of the problem. As O'Connell notes, what is now required is a more sophisticated understanding of arms that places weaponry into a more human context, no longer to be construed as oddities, alien instruments that have nothing important to say about human evolution, but rather as implements of profound importance. Indeed, weapons offer a window into many of humankind's worst fears and most compelling fantasies, and as such they should be regarded as instruments of great psychological significance.[3]

The claw and tooth, the horn and antler, the tusk and trunk are all weapons, weapons that predate humankind by eons upon eons. Indeed, life could not have struggled out of the swamp had it not been for weapons, both offensive and defensive. The immune system within the human body is in fact a "weapons system," a carefully choreographed response by tiny soldiers to foreign invasion. Use of the extended weapon vastly predates the human experience. As Robert Ardrey points out,

> In 1949 [Raymond] Dart dropped the other shoe. He had published a paper in the *American Journal of Physical Anthropology* claiming that *Australopithecus africanus* had gone armed. Study of some fifty-odd baboon skulls from various sites associated with the southern ape had revealed a curious, characteristic double depression. Dart concluded that the baboons had met sudden death at the hands of the southern ape: that the man-ape had used a weapon and that his favorite weapon had been the antelope humerus bone.[4]

Dart's findings made it clear that weapons, and the death-dealing due to them, had well preceded humankind, but in reality the destruction inflicted by the southern ape was different only in form. Animals had hunted, fought, and defended themselves since the dawn of time, and the southern ape's contribution was one simply of capacity, not result. Yet that expanded capacity would prove significant. Thus, long before the arrival of *Homo sapiens sapiens*, the line of protohuman ascent had already devised for him the extended weapon, a killing tool beyond his own limited physical means that could be wielded to advantage. Surely it was a product of greater cognition, this first weapon, the greater brain power of the southern ape, and it represented a marriage that would be expanded upon exponentially with the exponential increase in cognition of early humans. Still, none of these weapons argue for war, have caused wars, and should not, therefore, be confused with the central issue of this inquiry. The fact that weapons predated war by millions and millions of years successfully argues the case that weapons alone could not have caused war.

Yet the ancient link, indeed the ancestral link, of humankind and the weapon cannot be lightly dismissed. In fact, one evolutionary consideration is rather ominous in its implications. Robert Ardery, with whom I often disagree but whom I admire for his willingness to face even the most troubling question head-on, explains: "Weapons preceded man. Whether man is in fact a biological invention evolved to suit the purposes of the weapon must be a matter of future debate."[5] Put simply, Ardrey asks whether humans evolved weapons, or whether weapons somehow evolved humankind as a means to their (weapons) own development. The question may at first seem facetious, but serious consideration proves troubling, and the implications cannot be so easily brushed aside. Ardrey goes on:

> Man takes deeper delight in his weapons than in his women. He will pledge a treasury to one; a pittance to the other. From handaxe to hydrogen bomb his best efforts have been spent on the weapons perfection.... Ben Franklin first spoke of man as the tool-making animal. Now the British Museum publishes Kenneth Oakley's authoritative handbook on anthropology, *Man the Tool-Maker*, in which that definition is accepted. Any inspection of Oakley's handbook, however, should reveal that the continuity of development in man's cultural efforts is not truly that of the tool; it is that of the weapon.... Yet we dare not say so.[6]

Yet Robert O'Connell seems to disagree with Ardrey, if even but slightly. "Throughout most of human history the rate of weapons innovation has been very slow," notes O'Connell. "So slow, in fact, that we sometime have trouble comprehending it."[7]

The fact of the matter is that until the industrial revolution — inaugurated itself by the rise of the level four thought process — the evolution of weaponry had been very slow, and not the central focus of human industry that Ardrey insists it was. It has been this most recent burst in weapons development — over the last 180 years — that has impressed itself upon our conceptions, forcing the modern mind to think in terms of a constant arms race, when in fact for most of human history there had been only minor advances. It was not until the industrial revolution that the rate of arms innovation began to truly accelerate, and not until the world was seen through the eyes of a potential nuclear conflagration that the process seems to have careened entirely out of control.[8] In other words, while weapons have most certainly been a constant in humankind's long ascent, not until recently did the rate of their development increase exponentially; and it is that rapid increase that has branded our impressions with the specter of impending doom, of a world gone mad, and the weapon as the culprit.

Nowhere has this distortion been more keenly felt than in the rising halls of late level four and early level five consciousness, where the weapon itself (and virtually anything that smacks of war) has become a target of almost hysterical contempt, and any instrument of potential violence raged against. Pacifists and peace activists, as an example, constantly equate a reduction in arms to a reduction in the probabilities of warfare where generally no correlation can be demonstrated to exist. Blood (or red paint) is hurled at military weapons, access to military installations blocked, even the active status of ROTC units on college campuses railed against, as if the human impulse to war were something so insignificant it could be dulled or even thwarted entirely by acts of theatrical displeasure alone. Get rid of the weapons — or so, at least, the thinking goes — and war will evaporate like a bad dream. Unfortunately, neither war nor weapon are as simple as that.

That these demonstrations often border on the religions in terms of their zeal and focus should come as no surprise. Ken Wilber has pointed out that the religious and iconic symbols of an old psychological level generally become the detested, vilified symbols of the new.[9] In just this manner, for instance, the serpent, which was a primal symbol of power during the level two epoch, rapidly descended into villainy with the arrival of level three and the rise of the hero myth. It was the serpent, it can be recalled, that often guarded the ego and had to be defeated by the hero during the course of his trial, the serpent that represented the living, evil presence of the old order.[10] Recall as well, that it was Satan, having assumed the guise of a serpent, who tempted Eve in the Garden of Eden, a story that has seemingly linked evil with the serpent for eternity.

In the halls of modern academia, however, Satan has been replaced by the uniform, the tank, the rifle, and the grenade, and here they are despised with almost religious intensity. These now are the iconic symbols of the old order, and these symbols have taken on the same satanic qualities as the ancient serpent. As such, they are denied validity and subjected to ridicule and scorn, indeed, defeated by the new hero of the new order: the compassionate pacifist who hurls not spears but flowers. That this is simply act two, or perhaps three, of a very old play occurs to almost no one.

And while these demonstrations amount generally to little more than theater, they do in fact represent something new and, for humankind, something positive as well. They are the first, somewhat immature emergence of a new level of compassion and morality, a morality that can no longer accept war as a legitimate feature of the status quo, and a morality that places compassion above power. That this level has been initially fixated upon weapons as a target is understandable, even if misbegotten, for weapons are, in the end, little more than what men and women will make of them.

Weapons are, despite their eons of use and all the psychological cachet they have acquired as a result, in the final analysis simply what the dictionary defines them as, *tools*, extensions of humankind's meager physical abilities amplified by our robust intellectual and psychological endowment. The modern aircraft carrier and its attack group can, with its fleet of helicopters and remarkable communications capabilities, for instance, deliver troops, equipment, ordnance, and supplies to the modern battlefield with speed and precision. Yet it can also deliver food, medicine, shelter, and the appropriate personnel to a flood or earthquake-ravaged landscape with equal felicity, and this we have seen accomplished on more than one occasion. The choice is entirely in the hands of the current crop of men and women who sustain and command the vessels, and of those in political authority who control them. No ancient weapons archetype controls or guides this process. *Homo sapiens sapiens* is not the evolutionary tool of the weapon. Robert Ardrey was entirely correct in noting that the tools featured in Kenneth Oakley's handbook titled *Man the Tool-Maker* were fundamentally weapons, but he failed to take note of the equally obvious fact that those same weapons are, in the end, simply another form of tool. The antelope humerus bone raised as a club by the southern ape millions of years ago gave life not only to clubs but also to hammers, one to destroy, the other to build. Which it will be is decided by human beings alone, not some genetic or instinctual power beyond human knowledge or being. Indeed, were it the other way around it would relieve humankind of its failed responsibilities, responsibilities that loom ever larger as humanity enters the twenty-first century.

Weapons, demystified, return to their most basic, unglamorous meaning, the tool which Webster defined simply as "any instrument or device used for attack or defense in a fight or in combat." In that sense it can be understood that today their design and use rests entirely in the hands of human beings. From a psychological perspective — and weapons still hold immeasurable psychological power — weapons are simply a technical manifestation of mankind's immature fears, some legitimate, others fanciful, but either way, fears that have subconsciously dominated mankind's actions since our arrival upon this planet. The more fear the more weapons, their ingenious killing capacity having increased in conformity with humankind's developmental growth: the higher the level, the more evolved the weapon. Still, it is fear that drives the growth and destructive capabilities of weapons. Yet, as humans have matured as a subspecies, those irrational fears have begun to subside, and humans have become more balanced, more logical, less moved by irrational prejudice. That is surely the good news because it clearly implies that men and women of good conscience can alter the current state of affairs. Yet there is also little question that in our past many of those decisions have often been horrid, and that if humankind continues blindly down that same worn path, then we may all soon meet our doom. Still, there is significant hope that in the future those decisions will be different, far better in fact. Those reasons for this hope will be covered in detail in the next chapter.

Is it possible, then, that someday soon we will all share a world absent weapons, not only their physical presence but also the profound psychological fear they induce? Will they be lost, buried, burned, or beaten into plowshares? I think not. To begin with, we still live in a dangerous world, a world prone to violence, and while the trend toward level four/five continues across the globe, in the end there are no guarantees. This should cause no fear, however, for weapons themselves have never really been the problem. Rather the problem has been humankind's irrational needs, the deep-seated psychological flaws and projected drives that have been at the core of mankind's miseries since *Homo sapiens sapiens* first grasped the imminent demise of their fragile egos and lashed out in ungovernable protest. Human development will cause naturally the weapon to be reevaluated, reduced, and ultimately used with increasing compassion, reason, and constraint — to help the needy, defend the weak, and enforce just laws across the planet. None of this will happen overnight, of course, and none of it, as is typical of humankind's ragged growth, will be accomplished without error.

Ultimately weapons will not be banned, burned or beaten into plowshares. Surely, a different prospect awaits such a formidable partner of the human ascent. Already that fate has been decided, I think, and slowly, quietly, the weapon's future resting place already constructed by those few thoughtful souls in our culture who see to such things. These resting places are not cemeteries nor are they cathedrals, places of scorn or places of worship. Rather they are simple museums, respectful places where the artifacts of human evolution remain on quiet display for all to see. Here rest the Henry, Springfield, Winchester, and Sharpes rifles. On display are row upon row of pistols, of machine guns, of tanks, flintlocks, and artillery pieces. Here they rest in peace, their deadly service at long last complete, no more frightening on display than are the forks, spoons, and tableware of earlier generations. They await only the last generation of twenty-first century weapons to join them, artifacts that some future generation might then ponder in quiet reflection. Peacefully on display, they will be seen then for what they were, simply the death-dealing gadgets of a long-ago frightened beast, who, having grown up and away from the wars that once consumed him, requires them no more.

Eighteen

Peace

We humans are not quite as advanced as we like to think. We imagine that we have long outgrown the rough times and hard ways of our collective past, and that as modern individuals we now function at a much higher level than those who came before us. But that's not entirely true. The beliefs and superstitions of our long forgotten ancestors still haunt our language, thoughts, and actions, most often without even the slightest awareness on our part. Voodoo, good luck charms, holy relics, irrational prejudice, and other odd spells are still with us today as much as are our modern science, laborsaving devices, and supposedly rational politics. In many, many ways, our past still holds the present firmly in its grip.

War, too, is a product of this human immaturity, and that is why war remains with us still. The Homeric ethos, for instance — despite its increasing obsolescence due to the lethality of the modern battlefield — still drives many codes of athletic, legal, and business conduct, and it certainly reverberates throughout our modern military culture. We have a long way to go, we humans, simply to collectively reach the level four plateau. So if we ever hope to outgrow war it is imperative that we, as a subspecies, be able to distinguish forward from backward. In other words, we need to clearly grasp what behaviors constitute development as opposed to those that constitute regression, for, as our inquiry amply demonstrates, going backward has more than once been confused with human progress.

While the United States and the Soviet Union, fortunately, never came to nuclear blows during the cold war period, the post–World War II era was hardly a respite from more conventional warfare. "Between the years 1945 and 1990, while the fear of nuclear retaliation dissuaded the United States and the Soviet Union — and their respective European allies — from directly warring against each other, the rest of the world enjoyed no respite from conflict."[1] Indeed, over eighty wars were waged during this period, adding some twenty million lives to the list of those lost to twentieth century warfare.[2] These were primarily wars of national liberation (early level four), but also many others simply renewed ancient, tribal, and sectarian hostilities (early and late level three). Thus, for the conduct of war during the late twentieth century, it was more or less business as usual, only the methodologies and the extraordinary human toll having changed.

Today there are in essence only pockets of levels four and five populations across the globe, and while those populations are dwarfed by the remaining level three populations, they have over the past 150 years produced virtually all of the science we take for granted and that underlies — in the West, at least — our modern way of life. Indeed, so dramatic has been this developmental burst that the psychiatrist Clifford Anderson noted that "over the past century we have witnessed more collective movement along our maturational path than has been seen over the entire previous span of human history,"[3] an incredible development and a remarkable story in itself. But there is a troublesome side to this rapid and unprecedented growth.

In short, although anyone today could can purchase and use a microwave, cell phone, digital camera, or air conditioner, very few societies can actually design, manufacture, and distribute them. Likewise, while most level three populations are incapable of designing and manufacturing advanced weapons systems, almost any third- or fourth-world nation can buy and use them. And one of the sad, morally deficient stories of the twentieth century consists of just how many advanced, level four societies have been content to sell or even give away these systems to regimes that either did not safeguard or could not safeguard the weapons they received.

Thus a world rapidly "shrinking" as a result of level four technology is increasingly under threat by means of that same technology used against it by minds that could never produce the technology independently but are happy to acquire it then use it to their own ends. The horrific damage and loss of life associated with 9/11, for instance, is a result of profoundly ethnocentric, level three minds using the advanced technology of the level four world against itself. Long before that fateful day, Ken Wilber grasped the looming horror with startling clarity:

> For example, the atomic bomb is the product of formal-operational thinking (orange) [level 4], but once it exists, it can be used by individuals at lower levels of development, even though those levels could not themselves produce the bomb.... Stated in more general terms, one of humanity's constant nightmares has been that technological growth ... has always run ahead of ... growth in wisdom, care, and compassionate use of that technology.[4]

As humanity enters the twenty-first century the looming, and most prominent, international confrontation appears to be between the Western democracies and Islamic fundamentalism. This is a classic level three–level four confrontation, yet with a few new and curious twists. Those twists are principally the extreme ethnocentric stance of this most vocal, radical form of Islam — a stance that literally applauds the murder of any non–Islamic individual (infidel) in the name of God — and the international stage upon which the drama is being played out. This murderous injunction is *not*, however, representative of traditional Islamic religion, but rather a radical theological stance staked out by only the most desperate minds among a small percentage of fanatics. So there is at least some rational hope that more moderate voices will ultimately prevail within the Islamic community.

The Islamic religion is practiced around the world these days, and there are so many sects, ancient tribes, and interpretations alive and well that it is impossible to predict with any confidence how this confrontation might ultimately play itself out. No one has a crystal ball. Level three religion, which is generally quite dogmatic and ethnocentric, as pointed out in chapter fifteen, has over the course of history been a troubling instigator of wars, and that trend might well continue into the next century. If true, however, that trend would certainly have to buck the whole tide of human development that is clearly moving toward levels four, five, and beyond in the West. Indeed, while this development serves to isolate and infuriate Islamic militants all that much more, I suspect this drift toward war will not occur.

Information today travels at such an extraordinary rate and in such abundance that the sheer material benefits alone of an open level four society will in all probability soon impress themselves upon the Islamic majority. And ultimately people are remarkably alike — given the facts, they want to live in peace and raise their children to a better world. They want progress, and the very notion of progress is itself essentially an idea spawned by the level four capacity to imagine what might be.

Ultimately, extreme Islamic fundamentalism will be doomed in exactly the same manner as was the former Soviet Union — by the flimsiness of its own delusions. In the long run, I suspect most Muslims will come to understand that the confrontation is really not religious at all. Buddhism, Christianity, and Hinduism, after all, have all survived the evolutionary leap from level three to level four intact, and Islam can easily accomplish the same.

In that sense, the confrontation is not really about the *rise* of Islam at all, but rather the imminent *collapse* and failure of a very narrowly defined level three Mideastern worldview, and the ascension of a truly global level four society. *That* is precisely what terrifies Islamic terrorists. Terrorism, after all, is not the tool of choice of successful worldviews, but rather a tool of sheer desperation. As Wilber points out, any level "reacts negatively if challenged; it lashes out, using its own tools, whenever it is threatened."[5] And Islamic terrorism, most fundamentally understood, is the hysterical reaction of a level sensing its own, imminent demise.

This is why Islamic extremists hate and condemn everything Western, from Christianity to democracy itself, and categorize any Westerner as a crusader, because that is the literal time frame and worldview in which they are embedded. The Islamic extremists are not at war specifically with the United States, France, or Great Britain per se. They are at war with the entire level four experience because level four, like a vast, gleaming mirror, reflects the limitations and failures of their own worldview, and truths like that are not easily accepted. It is not mere coincidence that in Afghanistan, for instance, the Western powers labor to build schools only to see them promptly blown up by the Taliban. Schooling, after all, is the pathway to level four consciousness, and the Taliban, almost instinctively, appears to grasp the threatening potential of their presence. Most people in the Middle East understand it is time to grow, while some few would prefer to simply destroy the mirror, and the means to a better day. Stanford historian Victor Davis Hanson captures this unfocused anger in the words of the 9/11 terrorists themselves: "Instead of having any precise claim against America, these killers showed that their hurt arose from their own sense of envy and collective failure — as the now all-too-familiar references to 'being humiliated' and lost honor in their tape attest."[6]

The track record for lower levels prevailing against higher ones is not good, and I suspect the Middle East will slowly, awkwardly, and no doubt violently move toward democracy over the course of the next century. This does not, of course, diminish the potential catastrophic violence that might be inflicted by even a handful of dedicated terrorists, and this violent threat is one that, for the time being at least, must be taken very seriously. In the West this conflict has naturally been greeted with calls for healing and dialogue, admirable level four and five concepts that have absolutely no resonance in the Islamic, level three world, and as such only tend to confuse the issue. Remember, the instinctive level four/five desire to avoid conflict through discussion and compromise are only viewed as signs of weakness by the level three mind, and level three has only contempt for weakness.

Thus the failure of Islamic extremists to grasp just what it is they are contending with has so far been almost perfectly mirrored in the West, the invasion of Iraq providing only the most glaring example. While dispatching the inferior Iraqi military with relative ease, the Western occupation of the country proved another matter entirely. Americans expected to be greeted as liberators in Baghdad just as they had been in Paris in 1944, France being a solid level four nation at the time of the D-Day invasion. Instead they stumbled headfirst into a level three nightmare for which they were entirely unprepared.

Moreover, healing, as often prescribed by the postmodern level five mind for resolving every sort of conflict imaginable from war to racial discord, is by definition something that restores a system to its former health, harmony, and equilibrium. Since peace and racial harmony, as examples, have never historically prevailed on earth, healing represents an entirely inappropriate instrument for those positive changes. How, exactly, can a system be restored to a condition that never existed to begin with? How might we heal ourselves to a worldview that we never had? We can't! But we *can* grow toward it.

Global conflict will dissipate naturally once the elementary conditions that foster human development are widely in place, while in their absence war will continue to consume us all. Growth, not healing, will provide the instrument that finally ends the scourge of war. Human development, already accelerating to remarkable levels in many areas across the globe, is the one true, scientifically established path to the eradication of war and a better life for *all* humanity.

War arose some twelve thousand years ago as a result of human immaturity, and it reached a crescendo in the twentieth century with the staggering loss of perhaps ninety million lives. But the advent of nuclear weapons has made war both intellectually and operationally obsolete — at least in its most extreme, nuclear form, and the rise of level four consciousness has begun to create the conditions worldwide that will ultimately lead to the end of war as we have sadly experienced it for almost 12,000 years. What are those conditions?

Very roughly speaking, human beings develop in four distinct ways — physically, emotionally, mentally, and spiritually — and for proper development, all four categories must be addressed. While a detailed elaboration of all the elements necessary for all the various levels, is far beyond the scope of this book, here, at least, is a rough and partial outline of them.

The physical conditions requisite for growth are generally well known. These include, among other things, a nutritious diet, safe, comfortable housing, medical care, adequate sleep and exercise, and so on. These, naturally, are the most basic components of human growth.

Emotionally, humans require a loving, caring family or guardian to help identify, deal with, and master the various emotions and problems that emerge throughout life's circumstances. This requires great patience and maturity in and of itself on the part of the family; that patience allows the individual to initially discern right from wrong and then ultimately to understand and control their emotional responses. Mature individuals generally govern their emotions, while immature individuals are governed by their emotions.

Mentally the individual human requires access to comprehensive schooling and the necessary support to pursue that education over a lifetime. In the West this need has been instinctively met by the implementation of mandatory education from kindergarten through high school and an enormous expansion of college and university facilities.

Finally, individuals require the freedom to investigate and pursue — or not investigate and pursue — whichever spiritual path might appeal to them. A little understood fact regarding this is that the scientific West still tends to equate all spiritual approaches with primitive superstition or irrational dogma (levels two and three), however, just as each level has a different relation to war, so does each level also generate a different grasp of spirituality. While levels two and three spirituality might well appear superstitious and dogmatic to the rational mind (level four), there are also higher levels of development, and these levels have their own spiritual interpretations. Thus it is just as important to allow individuals to freely pursue their own notions of depth and meaning as it is to provide for their shelter, education, and emotional stability.

Over the past fifty years these conditions have begun to appear across the globe on a far greater scale than ever before, but they are still far from universal. Obviously, much needs to be done, and it may well take a century or more to tilt the human average from three to four. Yet as these basic conditions continue to emerge they will greatly accelerate the rate of human development. From that development will emerge a new chorus of reasonable voices, voices for which dialogue and compromise will seem the only sensible avenue for problem solving, and with that process war fighting will begin to recede from the human imagination. This will not happen overnight, but it *can* happen.

The end of war is hardly guaranteed, but as humanity enters the twenty-first century that prospect has moved now from a once fond but quixotic dream to at least a realistic possibility. That possibility rests squarely on the fact that war can now be understood, and as Conrad Lorenz noted in the prologue of this inquiry, wherever humankind has been able to discern the cause or causes of various phenomena we have been able to guide those factors to our benefit. And now that war can be understood, it can be guided, dealt with, and, we can be hopeful, eradicated.

The fact of the matter is, that process has already begun, even if but haltingly. As John Keegan states in *A History of Warfare*, one need not believe that humankind is free of a genetic disposition toward warfare "in order to be impressed by the evidence that mankind, wherever it has the option, is distancing itself from the institution of war...." He is "impressed by the evidence," Keegan states emphatically: "War, it seems to me, after a lifetime of reading about the subject, mingling with men of war, visiting the sites of war and observing its effects, may well be ceasing to commend itself to human beings as a desirable or productive, let alone rational, means of reconciling their discontents."[7]

Eighteen. Peace

For over 12,000 years humankind has been plagued by the dreadful reality of war. Try as we might, it remains far beyond our meager capabilities to catalogue how many souls have been lost, lives ruined, and families shattered as a result of this catastrophic undertaking. The costs, in the end, are simply incalculable. For centuries it seemed as though war was simply a permanent fixture of the human experience, and to a certain extent it was — for those times and for those levels. But no longer.

Today, perhaps for the very first time, it is entirely possible to imagine a world population developed enough to finally consign war to history's dustbin, and that world is not necessarily a distant dream at all. That possibility is now within reach, and because the process of growth can be understood, the task that looms before humanity is entirely achievable. While today much of the world's population remains at level three, many governments are functioning close to or at level four. To tip the balance fully toward democracy will surely require a great deal of toil, but what greater reward could there possibly be than to leave a world behind for tomorrow's children that is finally free of the scourge of war?

Recently Francis Fukuyama proclaimed in *The End of History and the Last Man* that democracy (level four) had in essence won history's marathon, delivering humankind therefore to the end of historical progression. What Fukuyama grasped clearly was the obvious advantages of level four development, economics, and government (in comparison to all lower levels), but what he failed to understand was that beyond level four there awaits a whole uncharted domain of progressive human potential. That potential represents an existence for *Homo sapiens sapiens* possibly far beyond our fondest hopes and dreams.

But that world will remain unclaimed unless the urge to war can finally be put aside, and that can only be accomplished with the establishment of more constructive conditions on a worldwide basis. That, today, is the challenge, but it is a challenge that is certainly achievable. And while the task of implementing those constructive conditions certainly stretches across the globe, it begins in each and every home. Wouldn't it be wonderful to think that in some far distant historical record our current generation, and perhaps the next few to follow, will be thankfully credited with the one accomplishment that will have made all future progress possible, quite possibly the greatest human accomplishment of all: *peace.*

Chapter Notes

Preface

1. John Keegan, *A History of Warfare* (New York: Alfred Knof, 1994), 3.

Prologue

1. Tony Allan, *Secrets of the Ancient Dead* (London: Duncan Baird, 2004), 104.
2. George Constable, ed., *The Human Dawn* (Alexandria, VA: Time-Life Books, 1990), 95.
3. Allan, 127.
4. Ibid..
5. John Keegan, *A History of Warfare* (New York: Vintage, 1993), 118.
6. Ibid., 117.
7. Ibid., as quoted, 117–118.
8. Ken Wilber, *One Taste* (Boston: Shambhala, 2000), as quoted, 306.
9. Robert L. O'Connell, *Of Arms and Men* (New York: Oxford University Press, 1989), 25.
10. Stanton A. Coblentz, *From Arrow to Atomic Bomb* (South Brunswick, NJ: Perpetua, 1967), 15.
11. Keegan, 120.
12. O'Connell, 23.
13. Ken Wilber, *Up from Eden* (Wheaton, IL: Quest Books, 1996), 109–110, 187.
14. Keegan, 118.
15. O'Connell, 27.
16. Ibid., 30.
17. Robert Ardrey, *The Social Contract* (New York: Dell, 1970), 257.
18. Keegan, 91.
19. O'Connell, 35.
20. Keegan, 140–141.
21. Ibid., 139.
22. Trevor N. Dupuy, *The Evolution of Weapons and Warfare* (New York, Da Capo, 1984), 2.
23. Ibid., 3–10.
24. O'Connell, 31–32.
25. Robert Ardrey, *The Territorial Imperative* (New York: Atheneum, 1966), 335.
26. Wilber, *Up from Eden*, 163–166.
27. Dupuy, 1.
28. Konrad Lorenz, *On Aggression* (New York: Bantam, 1966), 27.
29. Alfred North Whitehead, *Science and the Modern World* (New York: Macmillan, 1953), 187.

Chapter One

1. Malcolm S. Gordon, *Animal Physiology — Principles and Adaptations* (New York: Macmillan, 1977), 508.
2. Robert L. O'Connell, *Of Arms and Men* (New York: Oxford University Press, 1989), 28.
3. John Keegan, *A History of Warfare* (New York: Alfred A. Knopf, 1994), 79–80.
4. O'Connell, 26.
5. John Horgan, "Quitting the Hominid Fight Club: The Evidence Is Flimsy for Innate Chimpanzee — Let Alone Human — Warfare," Cross-check, Critical Views of Science in the News, June 29, 2010, accessed June 18, 2011. (http://www.scientificamerican.com/blog/post.cfm?id=quitting-the-hominid-fight-club-the-2010-06-29).
6. John Mitani, David P. Watts, and Sylvia J. Amsler, "Lethal Intergroup Aggression Leads to Territorial Expansion in Wild

Chimpanzees," *Current Biology*, Volume 20, Issue 12, R507-R508, 22 June 2010, 1.
7. Ibid., 1–2.
8. Ibid., 2.
9. Ibid.
10. Ibid.
11. Ibid., 1.
12. John Horgan, 1–2.
13. Ibid., 2.
14. Ibid., 3.
15. Ibid., as quoted, 3.
16. Ibid., 3.
17. Robert Ardrey, *African Genesis* (New York: Dell, 1961), 174.
18. Ibid.
19. Ibid., 354.
20. Ibid.
21. George Constable, ed., *The Human Dawn* (Alexandria, VA: Time-Life Books, 1987), 149–150.
22. George Constable, ed., *The Age of God-Kings* (Alexandria, VA: Time-Life Books, 1987), 12.
23. Constable, ed., *Human Dawn*, 137.
24. Edward O. Wilson, *Consilience, the Unity of Knowledge* (New York: Vintage, 1999), 315.
25. Keegan, 80.
26. Stanton Coblentz, *From Arrow to Atom Bomb* (South Brunswick, NJ: Perpetua, 1967), 10.
27. Ibid., 462.
28. Dr. Laurence J. Peter, *Peter's Quotations* (New York: Bantam Books, 1977), 517.
29. Dean G. Pruitt and Richard C. Snyder, *Theory & Research on the Causes of War* (Englewood Cliffs, NJ: Prentice-Hall, 1969), 5–14.

Chapter Two

1. Ken Wilber, *Sex, Ecology, Spirituality* (Boston: Shambhala, 1995), 153–204.
2. Carter Phipps, "Integral Politics Comes of Age," *What Is Enlightenment*, Oct.–Dec. 2007, Issue 38, 68.
3. Ken Wilber, *Up from Eden* (Wheaton, IL: Quest Books, 1996), 26–40.
4. F. Philip Rice, *The Adolescent—Development, Relationships, and Culture* (Boston: Allyn & Bacon, 1996), 37.

5. Jim Stempel, *When Beliefs Fail* (West Chester, PA: Chrysalis Books, 2001), 32.
6. Wilber, *Sex, Ecology, Spirituality*, 211.
7. Stempel, 47.
8. Ibid., 32.
9. Wilber, *Sex, Ecology, Spirituality*, 212.
10. Ibid., as quoted, 212.
11. Stempel, 48.
12. Wilber, *Sex, Ecology, Spirituality*, 218–219.
13. Ibid., 224.
14. Stempel, 50.
15. Wilber, *Sex, Ecology, Spirituality*, 231.
16. Stempel, 51.
17. Ibid.
18. Clifford Anderson, *The Stages of Life* (New York: The Atlantic Monthly Press, 1995), 104.
19. Ibid., 124.
20. Ibid., 134.
21. Rice, 38.
22. Anderson, 124.
23. Wilber, *Up from Eden*, 46.
24. Ibid.
25. M. Scott Momaday, "I Am Alive," *The World of the America Indian* (Washington, DC: National Geographic Society, 1979), 23.
26. Wilber, *Up from Eden*, 45.
27. Ibid., 51–52.
28. Ibid., as quoted, 50.
29. Stanton A. Coblentz, *From Arrow to Atom Bomb* (South Brunswick, NJ: Perpetua, 1967), as quoted, 77.
30. A.T. Olmstead, *History of Assyria* (New York: Charles Scribner's Sons, 1923), as quoted, 97.
31. Thomas H. Flaherty, ed., *The Enterprise of War* (Alexandria, VA: Time-Life Books, 1991), as quoted, 17.
32. *Holy Bible* (Carol Stream, IL: Tyndale House, 1992), 197.
33. Ken Wilber, *Up from Eden*, 93.

Chapter Three

1. W.T. Jones, *Masters of Political Thought, Machiavelli to Bentham* (Boston: Houghton Mifflin, 1968), 257.
2. *Webster's American Family Dictionary* (New York: Random House, 1998), 175.

3. Ken Wilber, *Up from Eden* (Wheaton, IL: Quest Books, 1996), 109.
4. George Constable, ed., *The Human Dawn* (Alexandria, VA: Time-Life Books, 1990), 95–96.
5. Ibid., 96.
6. Ibid., 99.
7. Wilber, *Up from Eden*, 93–95.
8. Ibid., 94.
9. Ibid., 180.
10. Ibid., as quoted, 96.
11. Joseph Campbell, *Myths to Live By* (New York: Bantam, 1972), 176.
12. Constable, ed., *The Human Dawn*, 141.

Chapter Four

1. Joseph Campbell, *Myths to Live By* (New York: Bantam, 1972), 23–24.
2. Carl Sagan, *The Dragons of Eden* (New York: Random House, 1977), 93–94.
3. Ken Wilber, *One Taste* (Boston: Shambhala, 2000), 305–318.
4. Ibid., 306.
5. Robert L. O'Connell, *Of Arms and Men* (New York: Oxford University Press, 1989), 124.
6. Jim Stempel, *When Beliefs Fail* (West Chester, PA: Chrysalis Books, 2001) 33–34.
7. Peter Green, *Alexander of Macedon* (Berkley, University of California Press, 1991), 59.
8. F. Philip Rice, *The Adolescent—Development, Relationships, and Culture* (Boston: Allyn & Bacon, 1996), 309–313.
9. Clifford Anderson, *The Stages of Life* (Boston: The Atlantic Monthly Press, 1995), 128.
10. Anthony Storr, *Human Aggression* (New York: Bantam, 1970), 59.

Chapter Five

1. Theodore Rubin, *Compassion and Self-Hate* (New York: Touchstone, 1975), 74; Anthony Storr, *Human Aggression* (New York: Bantam, 1970), 107–111.
2. Rubin, 74–75.
3. Storr, 106.
4. Ibid., 75.
5. Ibid., 77–78.
6. Jim Stempel, "The Dynamics of Spiritual Growth," *Concepts In Human Development*, Vol. 2, Issue 41, 15–16.
7. Storr, as quoted, 106.
8. Robert Ardrey, *The Social Contract* (New York: Delta Books, 1970) 257.
9. Stempel, "The Dynamics of Spiritual Growth," 15–16.
10. Rubin, 74–75; Jim Stempel, "Loving or Hating Yourself," *Concepts in Human Development*, Vol. 2, Issue 39, 19–20.
11. Zelda G. Knight, "Some Thoughts on the Psychological Roots of the Behavior of Serial Killers as Narcissists: An Objective Relations Perspective," *Social Behavior and Personality*, 34 (10), 2006, 1190.
12. David V. Canter, Laurence J. Alison, Emily Alison, and Natalia Wentink, "The Organized/Disorganized Typology of Serial Murder Myth or Model," *Psychology, Public Policy, and Law*, Vol. 10, No. 3, 2004, 302.
13. Knight, 1191.
14. Rubin, 74.
15. Knight, 1191.
16. Ibid., 74–75.
17. Ibid., 75.
18. Laurence J. Peter, *Peter's Quotations* (New York: Bantam, 1977), 517.
19. Ken Wilber, *Up from Eden* (Wheaton, IL: Quest Books, 1996), as quoted, 208.
20. Rush W. Dozier, Jr. *Fear Itself* (New York: St. Martin's Press, 1998) 150.
21. Wilber, *Up from Eden*, 121.
22. Ibid., 159.
23. Ibid.
24. Ibid., 161.
25. A.T. Olmstead, *History of Assyria* (New York: Charles Scribner's Sons, 1923), as quoted, 97.
26. Robert L. O'Connell, *Of Arms and Men* (New York: Oxford University Press, 1989), 43.
27. Stanton A. Coblentz, *From Arrow to Atom Bomb* (South Brunswick, NJ: Perpetua,1967), 61.
28. Wilber, *Up from Eden*, as quoted, 297.
29. Chris Hedges, *War Is a Force That Gives Us Meaning* (New York: Anchor Books, 2003), 144.
30. Wilber, *Up from Eden*, 164.

31. Ibid., as quoted, 298.
32. Ibid., 294.
33. O'Connell, 33.
34. Ibid.
35. Ibid.

Chapter Six

1. C.G. Jung, *The Basic Writings of C.G. Jung* (Princeton: Princeton University Press, 1990), 343.
2. Ibid., 343.
3. Ken Wilber, *Sex, Ecology, Spirituality* (Boston: Shambhala, 1995), 220.
4. Ibid., 220.
5. Ibid., 247–248.
6. Robert Robertson, *Jungian Psychology* (York Beach, ME: Nicholas-Hays, 1992), 43.
7. Robert Ardrey, *The Social Contract* (New York: Delta Books, 1971), 268.
8. John Keegan, *A History of Warfare* (New York: Alfred A. Knopf, 1994), 384–385.
9. Wilber, *Sex, Ecology, Spirituality*, 193–194.
10. Ibid., 194.
11. Robert L. O'Connell, *Of Arms and Men* (New York: Oxford University Press, 1989), 39.
12. Ibid., 40–41.
13. Ibid., 39.
14. Ibid., 41.
15. Ken Wilber, *Up from Eden* (Wheaton, IL: Quest Books, 1996), 188.
16. F. Philip Rice, *The Adolescent—Development, Relationships, and Culture* (Boston: Allyn & Bacon, 1996), 14.
17. Ibid., 150–155.
18. Wilber, *Up from Eden*, 199–209.
19. Ibid., 182.
20. Ibid., 179.
21. Anthony Storr, *Human Aggression* (New York: Bantam, 1970), 53.
22. Wilber, *Up from Eden*, 192.
23. Ibid., 192.
24. Ibid.
25. O'Connell, as quoted, 47.
26. Rice, 152–153.
27. Ibid., 152.
28. O'Connell, 46.
29. Ibid., 47–48.
30. Ibid., 49.
31. Stanton A. Coblentz, *From Arrow to Atom Bomb* (South Brunswick, NJ: Perpetua, 1967), 116.

Chapter Seven

1. Peter Green, *Alexander of Macedon* (Berkeley: University of California Press, 1991), 393.
2. Trevor N. Dupuy, *The Evolution of Weapons and Warfare* (New York: Da Capo, 1984), 30–33.
3. Thomas H. Flaherty, ed., *The Enterprise of War* (Alexandria, VA: Time-Life Books, 1991), 22.
4. Dupuy, 12.
5. Flaherty, ed., 22.
6. Ibid.
7. Stanton A. Coblentz, *From Arrow To Atom Bomb* (South Brunswick, NJ: Perpetua, 1967), 117.
8. Green, 407.
9. Coblentz, 117.
10. Ibid.
11. Ibid., 118.
12. Green, 396.
13. Ibid., 400.
14. Ibid., as quoted, p. 410.
15. Ibid., 474–477.
16. Ibid., 478.
17. Coblentz, 116.
18. Ken Wilber, *Up from Eden* (Wheaton, IL: Quest Books, 1996), as quoted 162.
19. Green, 487.
20. Ibid., 488.
21. Dupuy, 14.
22. Robert L. O'Connell, *Of Arms and Men* (New York: Oxford University Press, 1989), 187–237.
23. Ibid., 62.
24. Coblentz, 120.
25. Thomas Jefferson, *The Declaration of Independence*.
26. George Constable, ed., *Winds of Revolution* (Alexandria, VA: Time-Life Books, 1990), 110.

Chapter Eight

1. Roland Bainton, *Here I Stand: A Life of Martin Luther* (New York: Abingdon-Cokesbury Press, 1950), 181.

2. Ibid., 183.
3. George C. Constable, ed., *The European Emergence* (Alexandria, VA: Time-Life Books, 1989), 12.
4. Bainton, 17–36.
5. Ibid., 185.
6. Constable, ed., *The European Emergence*, 17.
7. Ken Wilber, *Sex, Ecology, Spirituality* (Boston: Shambhala, 1995), 178.
8. F. Philip Rice, *The Adolescent—Development, Relationships, and Culture* (Boston: Allyn & Bacon, 1996), 147.
9. Roger A. Johnson, ed., *Psychohistory and Religion* (Philadelphia: Fortress Press, 1977), 88.
10. Constable, ed., *Winds of Revolution* (Alexandria, VA: Time-Life Books, 1990), 8.
11. Ibid., 9.
12. Stanton A. Coblentz, *From Arrow to Atom Bomb* (South Brunswick, NJ: Perpetua, 1967), 294.
13. Robert L. O'Connell, *Of Arms and Men* (New York: Oxford University Press, 1989), 167.
14. Michael Stephenson, *Patriot Battles* (New York: HarperCollins, 2007), as quoted, 313.
15. O'Connell, 196.
16. Coblentz, 340.

Chapter Nine

1. Joseph J. Ellis, *His Excellency, George Washington* (New York: Vintage, 2005), 126.
2. David McCullough, *1776* (New York: Simon & Schuster, 2005).
3. Ibid., 25.
4. Ibid., 24.
5. Clifford Anderson, M.D., *The Stages of Life* (New York: The Atlantic Monthly Press, 1995), 134–135.
6. George Constable, ed., *Winds of Revolution* (Alexandria, VA: Time-Life Books, 1990), 98–99.
7. McCullough, 25.
8. Ibid., 35.
9. Ibid.
10. Ellis, 77.
11. Ibid., 74.
12. McCullough, 249.

13. Ibid., 267.
14. Ellis, 98.
15. McCullough, 273.
16. Ibid.
17. Chris Hedges, *War Is a Force That Gives Us Meaning* (New York: Anchor Books, 2003), 148.
18. George H. Sabine, *A History of Political Theory* (New York: Holt, Rinehart and Winston, 1961), 669.
19. Thomas Jefferson, *The Declaration of Independence*.
20. Ken Wilber, *Sex, Ecology, Spirituality* (Boston: Shambhala, 1995) 234.
21. The Constitution of the United States of America.
22. Ellis, as quoted, 78.
23. Sabine, 669.
24. Anderson, 119.
25. Ellis, 91.
26. Ibid., 189.
27. Ibid., 141–142.
28. Ibid., as quoted, 142–143.
29. Ibid., 141.
30. Ibid., as quoted, 139.
31. Ibid., 142.

Chapter Ten

1. *The Library of Congress Civil War Desk Reference* (New York: Simon & Schuster, 2002), 623.
2. Ibid., 330–331.
3. Ibid., 376–383.
4. Ibid., 376.
5. Ibid.
6. Ken Wilber, *One Taste* (Boston: Shambhala, 2000), 302.
7. *The Library of Congress Civil War Desk Reference* (New York: Simon & Schuster, 2002), 88.
8. Ibid.
9. Ibid.
10. Ibid.
11. Ibid., 89.
12. Ibid.
13. Ken Wilber, *One Taste* (Boston: Shambhala, 2000), 302.
14. W.M. Beauchamp, "Iroquois Women," *The Journal of American Folk-Lore*, Vol. XIII, April–June 1900, No. XLIX, 83.

15. Constable, ed., *Winds of Revolution* (Alexandria, VA: Time-Life Books, 1990), 123–124.
16. Edward Porter Alexander, *Fighting for the Confederacy* (Chapel Hill: University of North Carolina Press, 1989), 503.
17. Robert L. O'Connell, *Of Arms and Men* (New York: Oxford University Press, 1989), 197.
18. Ibid.
19. Ibid.
20. Ibid., 191.
21. Ibid.
22. Ernest B. Furguson, *Not War but Murder* (New York: Vintage, 2000), as quoted, 20.
23. *The Library of Congress Civil War Desk Reference* (New York: Simon & Schuster, 2002), 494.
24. O'Connell, 200.
25. Ivan Musicant, *Divided Waters: The Naval History of the Civil War* (Edison, NJ: Castle Books, 2000), 136–137.
26. Ibid., 141.
27. Ibid., 136.
28. Ibid., 153.
29. Ibid., 156.
30. Ibid.

Chapter Eleven

1. Joseph J. Ellis, *His Excellency, George Washington* (New York: Vintage Books, 2005), 201.
2. Ibid., 202.
3. Clifford Anderson, *The Stages of Life* (New York: The Atlantic Monthly Press, 1995), 121.
4. Ibid.
5. Gary Wills, *Lincoln at Gettysburg* (New York: Touchstone, 1992), 125–130.
6. Alpheus Thomas Mason and William M. Beaney, *American Constitutional Law* (Englewood Cliffs, NJ: Prentice-Hall, 1968), 34.
7. Stephen B. Oates, *With Malice Toward None* (New York: HarperPerennial, 1977) 253.
8. Paul M. Angle and Earl Schenck Miers, *The Living Lincoln* (New York, Barnes & Noble, 1955), 576.
9. Oates, 333.
10. Ibid., 325.
11. Catherine Ann Edmondston, *Journal of a Secesh Lady* (Raleigh, NC: Division of Archives and History, 1979), 315.
12. Ibid.
13. Oates, 363.
14. Wills, 24.
15. Ibid., as quoted, 263.
16. Ibid., 146–147.
17. Carl Sandburg, *Abraham Lincoln: The Prairie Years and the War Years* (New York: Dell, 1954) as quoted, 412.
18. Ibid., as quoted, 411.
19. Ibid., as quoted, 412.
20. Ibid., as quoted, 411.
21. Angle and Miers, 592.
22. A.T. Olmstead, *History of Assyria* (New York: Charles Scribner's Sons, 1923), as quoted, 97.
23. Angle and Miers, 640.
24. The Constitution of the United States of America, Amendment XIII.
25. Oates, 405.
26. Wills, 145.
27. Anderson, 128.
28. Oates, 431.

Chapter Twelve

1. Jason McManus, ed., *The World in Arms* (Alexandria, VA: Time-Life Books, 1989), 8.
2. Ibid., 17.
3. Ibid.
4. Robert L. O'Connell, *Of Arms and Men* (New York: Oxford University Press, 1989), 211.
5. John Keegan, *The First World War* (New York: Alfred Knopf, 2001), 19–21.
6. Barbara W. Tuchman, *The Guns of August* (New York: Ballantine, 1962), 26.
7. Stanton A. Coblentz, *From Arrow to Atom Bomb* (South Brunswick, NJ: Perpetua, 1967), 417.
8. Keegan, 67.
9. Ibid., 71.
10. Tuchman, 377.
11. Ibid., 45.
12. Ibid., as quoted, 46.
13. McManus, ed., *The World in Arms*, 22.
14. O'Connell, 246.
15. Ibid.

16. Ibid., 244.
17. Ibid., 242.
18. Ibid.
19. Ibid., 246.
20. Ibid., 242.
21. Ibid., 242–243.
22. "We Don't Have to Be 'Here Now,'" a conversation between Ken Wilber and Andrew Cohen, *What Is Enlightenment*, Issue 32, May 2006, 23. A conversation between Ken Wilber and Andrew Cohen.
23. McManus, ed., *The World in Arms*, 15.
24. Woodrow Wilson. "Transcript of President Wilson's Fourteen Points, 1918," *100 Milestone Documents*, National Archives, n.d. Web. 11 Jul 2011. <http://www.ourdocuments.gov/doc.php?=62&page=transcript>.

Chapter Thirteen

1. Robert L. O'Connell, *Of Arms and Men* (New York: Oxford University Press, 1989), 270.
2. Ibid., 274.
3. Stanton A. Coblentz, *From Arrow to Atomic Bomb* (South Brunswick, NJ: Perpetua, 1967), 418.
4. George Constable, ed., *Shadow of the Dictators* (Alexandria, VA: Time-Life Books, 1989), 28.
5. Coblentz, 420.
6. Constable, ed., *Shadow of the Dictators*, 30.
7. The Library of Congress, *World War II* (New York: Simon & Schuster, 2007), 18.
8. Constable, ed., *Shadow of the Dictators*, 27.
9. O'Connell, 280.
10. Ibid., 281.
11. Library of Congress, *World War II*, 691–692.
12. John Strauson, *Hitler as a Military Commander* (London: Batsford, 1971), as quoted, 213.
13. Coblentz, 432.
14. Constable, ed., *Shadow of the Dictators*, 35.
15. Ibid.
16. Ibid., 40.

17. Library of Congress, *World War II*, 428.
18. Ibid.
19. Ibid., 481.
20. Constable, ed., *Shadow of the Dictators*, 65.
21. Ibid., 69.
22. O'Connell, 283.
23. Ibid., 290–291.
24. Ibid., 291.
25. Ibid.
26. Ibid.
27. Coblentz, 429.
28. Library of Congess, *World War II*, xiii.
29. Geoffrey C. Ward and Ken Burns, *The War* (New York: Alfred A. Knopf, 2007), as quoted, 214.
30. O'Connell, 293.
31. Ibid., 286.
32. Ibid., 290.
33. Ibid., 294.
34. Ibid., as quoted, 294–295.
35. Ibid., 296.

Chapter Fourteen

1. Constable, ed., *The Enterprise of War* (Alexandria, VA: Time-Life Books, 1990), 163.
2. Robert O'Connell, *Of Arms and Men* (New York: Oxford University Press, 1989), 301.
3. Ken Wilber, *Sex, Ecology, Spirituality* (Boston: Shambhala, 1995), 195.
4. George H. Sabine, *A History of Political Theory* (New York: Holt, Rinehart and Winston, 1961), 755.
5. Ibid., 758.
6. Ibid., 760.
7. Ibid., 854.
8. Ibid.
9. Ibid., 870–873.
10. Ken Wilber, *A Theory of Everything* (Boston: Shambhala, 2000), 122.
11. O'Connell, 289.

Chapter Fifteen

1. Ken Wilber, *A Theory of Everything* (Boston: Shambhala, 2000), 9.
2. Ibid.

3. Ibid.
4. Ibid., 9–10.
5. Ibid., 105.
6. Ibid., 10.
7. Ibid., 10–11.
8. Ken Wilber, *One Taste* (Boston: Shambhala, 2000), 241.
9. Ken Wilber, *A Theory of Everything*, 11.

Chapter Sixteen

1. Ken Wilber, *A Theory of Everything* (Boston: Shambhala, 2000), 11.
2. Ibid., 88.
3. Ken Wilber, *Integral Psychology, Consciousness, Spirit, Psychology, Therapy* (Boston: Shambhala, 2000), 51.
4. Wilber, *A Theory of Everything*, 86–87.
5. Ibid., 85.
6. Wilber, *Integral Psychology*, 41–42.
7. Wilber, *A Theory of Everything*, 84.

Chapter Seventeen

1. *Webster's American Family Dictionary* (New York: Random House, 1998), 1057.
2. Robert L. O'Connell, *Of Arms and Men* (New York: Oxford University Press, 1989), 5.
3. Ibid.
4. Robert Ardrey, *African Genesis* (New York: Dell, 1961), 30.
5. Ibid., 318.
6. Ibid., 206–207.
7. O'Connell, 9–10.
8. Ibid., 9.
9. Ken Wilber, *Up from Eden* (Wheaton, IL: Quest Books, 1996), 192–193.
10. Ibid.

Chapter Eighteen

1. George Constable, ed., *The Enterprise of War* (Alexandria, VA: Time-Life Books, 1991), 164.
2. Ibid.
3. Clifford Anderson, M.D., *The Stages of Life* (New York: Atlantic Monthly Press, 1995), 133.
4. Ken Wilber, *A Theory of Everything* (Boston: Shambhala, 2000), 103.
5. Ken Wilber, *Integral Psychology* (Boston: Shambhala, 2000), 51.
6. Victor Davis Hanson, "Bin Laden Tape Shows We Have a Lot to Learn About Our Terrorist Foes," *The Baltimore Sun*, Sept. 15, 2006.
7. John Keegan, *The History of Warfare* (New York: Vintage, 1994), 59.

Bibliography

Alexander, Edward Porter. *Fighting for the Confederacy: The Personal Recollections of General Edward Porter Alexander.* Gary W. Gallagher, ed. Chapel Hill: University of North Carolina Press, 1989.

Allan, Tony. *Secrets of the Ancient Dead.* London: Duncan Baird, 2004

Anderson, Clifford, M.D. *The Stages of Life.* Boston: Atlantic Monthly Press, 1995.

Ardrey, Robert. *African Genesis.* New York: Dell, 1967.

———, *The Social Contract.* New York: Delta, 1971.

———, *The Territorial Imperative.* New York: Atheneum, 1966

Armstrong, Karen. *A History of God.* New York: Alfred Knopf, 1993.

Bainton, Roland, H. *Here I Stand: A Life of Martin Luther.* New York: Abingdon-Cokesbury Press, 1950.

Barrett, Correlli. *The Great War.* London: BBC, 1979.

Beauchamp, W.M. "Iroquois Women." *The Journal of American Folk-Lore*, Vol. XIII, No. XLIX, April–June 1900.

Booth, Owen, and John Walton. *The Illustrated History of World War II.* Edison, NJ: Chartwell, 1998.

Boritt, Gabor S., ed. *The Gettysburg Nobody Knows.* New York: Oxford University Press, 1997.

Campbell, Joseph. *Myths to Live By.* New York: Bantam, 1974.

Canter, David, V., Laurence J. Alison, Emily Alison, and Natalia Wentink, "The Organized/Disorganized Typology Of Serial Murder, Myth or Model?" *Psychology, Public Policy and Law*, Vol. 10, No. 3, 2004.

Cartledge, Paul. *The Spartans: The World of the Warrior-Heroes of Ancient Greece.* Woodstock, NY: Overlook Press, 2003.

Coblentz, Stanton A. *From Arrow to Atomic Bomb.* South Brunswick, NJ: Perpetua, 1967.

Cohen, Andrew. "Conflict, Creativity and the Nature of God, Dialogue VI." *What Is Enlightenment*, Issue 26, Aug.–Oct. 2004.

———, "We Don't Have to 'Be Here Now.'" *What Is Enlightenment*, Issue 54, May 2006.

Constable, George, ed. *The European Emergence.* Alexandria, VA: Time-Life Books, 1989.

———, ed., *Shadow of the Dictators.* Alexandria, VA: Time-Life Books, 1989.

———, ed., *Winds of Revolution.* Alexandria, VA: Time-Life Books, 1990.

Corson, William R. *The Betrayal.* New York: Ace Books, 1968.

Crick, Francis. *The Astonishing Hypothesis.* New York: Touchstone, 1995.

Dann, John C., ed. *The Revolution Remembered.* Chicago: University of Chicago Press, 1980.

Dimock, Martha McHutchinson. *A Chronicle of the Revolutionary War.* New York: Perennial Library, 1976.

Dozier, Rush W., Jr. *Fear Itself.* New York: St. Martin's Press, 1998.

Dupuy, Trevor N. *The Evolution of Weapons and Warfare* New York: Da Capo, 1984.

Ellis, Joseph J. *His Excellency, George Washington.* New York: Vintage, 1994.

Ferling, John. *Almost a Miracle: The American Victory in the War of Independence.*

New York: Oxford University Press, 2007.

Flaherty, Thomas H., ed. *The Enterprise of War*. Alexandria, VA. Time-Life Books, 1991.

———, ed. *The Rise of Cities*. Alexandria, VA: Time-Life Books, 1991.

Furgurson, Ernest B. *Not War but Murder: Cold Harbor 1864*. New York: Vintage, 2000.

Gibbon, Edward. *The Decline and Fall of the Roman Empire*. New York: Dell, 1963.

Gordon, Malcom S. *Animal Physiology, Principles and Adaptations*, New York: Macmillan, 1977.

Grant, Ulysses S. *Personal Memoirs*. New York: Barnes & Noble, 2003.

Green, Peter. *Alexander of Macedon*. Berkeley: University of California Press, 1989.

Grunwald, Henry Anatole, Ed. *The Age of God-Kings*. Alexandria, VA: Time-Life Books, 1987.

Hampshire, Stuart. *The Age of Reason: The 17th Century Philosophers*. New York: Mentor, 1956.

Hanson, Victor Davis. "Bin Laden Tape Shows We Have a Lot to Relearn about Our Terrorist Foes." *The Baltimore Sun*, Sept. 15, 2006.

Hedges, Chris. *War Is a Force That Gives Us Meaning*. New York: Anchor Books, 2003.

Holmes, Oliver Wendell, Jr. *Touched with Fire: Civil War Letters and Diary of Oliver Wendell Holmes, Jr*. New York: Fordham University Press, 2000.

Homer. *The Odyssey*. W.H.D. Rouse, trans. New York: Mentor, 1937.

Horgan, John. "Quitting the Hominid Fight Club: The Evidence Is Flimsy for Innate Chimpanzee — Let Along Human — Warfare." Cross-check, Critical Views of Science in the News, June 29, 2010, Accessed June 18, 2011, *http://www.scientific american.com/blog/post.cfm?id-quitting-the-hominid-fight-club-the-2010-06-29*.

Ignatius, David. "The Future of Warfare." *The Washington Post*, January 2, 2011.

James, William. *Varieties of Religious Experience*. New York: Mentor, 1958.

Jaynes, Julian. *The Origin of Consciousness in the Breakdown of the Bicameral Mind*. Boston: Houghton Mifflin, 1970.

Johnson, Roger A., Ed. *Psychohistory and Religion*. Philadelphia: Fortress Press, 1977.

Jones, W.T. *Masters of Political Thought*. Boston: Houghton Mifflin, 1968.

Jung, C.G. *The Basic Writings of C.G. Jung*. Princeton: Princeton University Press, 1990.

Keegan, John. *The American Civil War*. New York: Alfred A. Knopf, 2009.

———. *The First World War*. New York: Alfred A. Knopf, 2001.

———. *A History of Warfare*. New York: Vintage, 1993.

Knight, Zelda G. "Some Thoughts on the Psychological Roots of the Behavior of Serial Killers as Narcissists: An Object Relations Perspective." *Social Behavior and Personality*, 34 (10) 2006.

Krycka, Kevin C. "Shamanic Practices and the Treatment of Life-threatening Medical Conditions." *The Journal of Transpersonal Psychology*, Vol. 32, November 2000.

Larson, Edward J. *Summer for the Gods*. New York: Basic Books, 1997.

Leslie, John, ed. *Modern Cosmology & Philosophy*. Amherst: Prometheus Books, 1998.

The Library of Congress. *World War II*. New York: Simon & Schuster, 2007.

Library of Congress Civil War Desk Reference. New York: Simon & Schuster, 2002.

Lorenz, Konrad. *On Aggression*. New York: Bantam, 1966.

Mascaro, Juan, trans. *The Bhagavad Gita*. New York: Penguin, 1962.

McCullough, David. *1776*. New York: Simon & Schuster, 2005.

McManus, Jason, Ed. *The World in Arms*. Alexandria, VA: Time-Life Books, 1989.

Mishlove, Jeffrey. *The Roots of Consciousness*. New York: Marlowe, 1975.

Mitani, John, David Watts, and Sylvia Amsler. "Lethal intergroup aggression leads to territorial expansion in wild chimpanzees," *Current Biology*, Volume 20, Issue 12, R507-R508, 22 June 2010.

Morison, Samuel Eliot. *The Two-Ocean War: A Short History of the United States Navy in the Second World War*. Boston: Back Bay Books, 1963.

Murphy, Michael. *The Future of the Body*. Los Angeles: Jeremy P. Tarcher, 1992.

Musicant, Ivan. *Divided Waters: The Naval History of the Civil War*. Edison, NJ: Castle Books, 2000.

National Geographic Society. *The World of the American Indian*. Washington, DC: National Geographic Society, 1979.

O'Connell, Robert L. *Of Arms and Men: A History of War, Weapons, and Aggression*. New York: Oxford University Press, 1989.

Olmstead, A.T. *History of Assyria*. New York: Charles Scribner's Sons, 1923.

Ouspensky, P.D. *The Psychology of Man's Possible Evolution*. New York: Vintage, 1974.

Phipps, Carter. "Integral Politics Comes of Age." *What Is Enlightenment*, Issue 38, Oct.–Dec. 2007.

———. "Is God a Pacifist?" *What Is Enlightenment*, Issue 26, Aug.–Oct. 2004.

———. "The Moral Dilemma of Pacifism in a World of War." *What Is Enlightenment*, Issue 26, Aug.–Oct. 2004.

Pratt, Fletcher. *A Short History of the Civil War*. New York: Pocket Books, 1961.

Pruitt, Dean G., and Richard C. Snyder. *Theory & Research on the Causes of War*. Englewood Cliffs, NJ: Prentice-Hall, 1969.

Rice, Philip F. *The Adolescent—Development, Relationships, and Culture*. Boston: Allyn & Bacon, 1996.

Ricks, Thomas E. *Fiasco: The American Military Adventure in Iraq*. New York: Penguin, 2006.

Robertson, Robert. *Jungian Psychology*. York Beach, ME: Nicolas-Hays, 1992.

Roemischer, Jessica. "The Never-Ending Upward Quest." *What Is Enlightenment*, Fall/Winter 2002.

Rubin, Throdore I., M.D. *Compassion and Self-Hate: An Alternative to Despair*, New York: Simon & Schuster, 1975.

Sabine, George H. *A History of Political Theory*. New York: Holt, Rinehart and Winston, 1961.

Sagan, Carl. *Broca's Brain: Reflections on the Romance of Science*. New York: Random House, 1974.

———. *The Dragons of Eden: Speculations on the Evolution of Human Intelligence*. New York: Random House, 1977.

Sandburg, Carl. *Abraham Lincoln: The Prairie Years and the War Years*, New York: Dell, 1954.

Searle, John. *Minds, Brains, and Science*, Cambridge: Harvard University Press, 1984.

Sears, Stephen W. *Gettysburg*. New York: Houghton Mifflin, 2003.

Skinner, B.F. *Science and Human Behavior*. New York: The Free Press, 1953.

Smith, Houston. *The Religions of Man*. New York: Harper & Row, 1986.

Stempel, Jim. *The Battle of Glendale: The Day the South Nearly Won the Civil War*. Jefferson, NC: McFarland, 2011.

———, "The Dynamics of Spiritual Growth." *Concepts in Human Development*, Vol. 2, Issue 41, 2001.

———, "Loving or Hating Yourself." *Concepts in Human Development*, Vol. 2, Issue 39, 2000.

———. *When Beliefs Fail: A Psychology of Hope*. West Chester, PA: Chrysalis Books, 2001.

Stephenson, Michael. *Patriot Battles: How the War of Independence Was Fought*. New York: Harper Collins, 2007.

Storr, Anthony. *Human Aggression*. New York: Bantam, 1970.

Strauson, John. *Hitler as a Military Commander*. London: Batsford, 1971.

Sun-tzu. *The Art of War*. John Minford, trans. New York: Viking, 2002.

Tuchman, Barbara W. *The Guns of August*. New York: Ballantine, 1962.

Wagner, Margaret E., Gary W. Gallagher, and Paul Finkelman, eds. *The Library of Congress Civil War Desk Reference*. New York: Simon & Schuster, 2002.

Ward, Geoffrey C., and Ken Burns, *The War*. New York: Alfred A. Knopf, 2007.

Whitehead, Alfred North. *Science and the Mordern World*. New York: Macmillan, 1953.

Wilber, Ken. *The Atman Project: A Transpersonal View of Human Development*. Wheaton, IL: Quest Books, 1980.

———. *Integral Psychology, Consciousness, Spirit, Psychology, Therapy*. Boston: Shambhala, 2000.

———. *One Taste: Daily Reflections on Integral Spirituality*. Boston: Shambhala, 2000.

———. *Sex, Ecology, Spirituality: The Spirit of Evolution*, Boston: Shambhala, 1995.

———. *The Simple Feeling of Being: Embracing Your True Nature.* Boston: Shambhala, 2004.

———. *A Theory of Everything: An Integral Vision for Business, Politics, Science and Spirituality.* Boston: Shambhala, 2000.

———. *Up from Eden: A Transpersonal View of Human Evolution.* Wheaton, IL: Quest Books, 1996.

Wilber, Ken, Jack Engler, and Daniel Brown, *Transformations of Consciousness.* Boston: Shambhala, 1986.

Wilson, Edward O. *Consilience: The Unity of Knowledge.* New York: Vintage, 1998.

Wilson, Woodrow. "Transcript of President Wilson's Fourteen Points. 1918." *100 Milestone Domcuments*, National Archives, n.d. Web. 11 Jul 2011 *http://www.ourdocuments.gov/doc.php?=62&page=transcript.*

Index

Achilles 68
Alexander, Gen. Edward Porter 117
Alexander the Great 71, 72–82
Amsler, Sylvia J. 13
Anderson, Clifford 27, 28, 38, 48, 94, 104, 126, 135, 192
archetype 62–71
Ardrey, Robert 8, 16–20, 48, 53, 64, 186, 187
Aristotle 47
Articles of Confederation 93
Ashurnasir-pal 33, 57, 134
atomic bomb 162

Blitzkreig 155
Booth, John Wilkes 136
Breed's Hill 93
Bronowski, Jacob 12, 18
Burgoyne, Gen. John 97

Campbell, Joseph 41, 43
Chamberlain, Neville 154–155
Coblentz, Stanton 4, 20, 58, 70–80, 87, 91, 140, 159
Cold War 163
College of William and Mary 96
Concord Bridge 87
concrete-operational 25
Continental Congress 93
Cornwallis, Gen. Charles 100
Cowpens 106

Darius III 73
Dart, Raymond 186
Declaration of Independence 81, 87, 93, 95, 102, 103, 126, 174
Demonic Male Theory 16

De Wall, Frans 15
Dupuy, Trevor 8, 73

Eden 44
Ellis, Joseph 98, 99, 106, 110
Emancipation Proclamation 127–130
Euripides 47
Everett, Edward 133

Fertile Crescent 3
formal-operational 26
Fourteen Points 149
Fukuyama, Francis 197
Fuller, J.F.C. 143

Galilei, Galileo 85
Gates, Horatio 106
Gaugamela 73
Gettysburg Address 132–133
Goodall, Jane 15
Granicus River 73
Green, Peter 3, 47, 74, 77, 78
Greene, Nathanael 105
Greenwood, John 98

Hannibal 78
Hapsburgs 138
Harvard College 96
Hedges, Chris 58, 102
Hegel, Georg 164
Henry Rifle 120
hero myth 68
Hitler, Adolf 89, 150, 151–162
Hobbes, Thomas 4
Homer 68–91, 112, 117, 120–124, 143, 144, 171, 191
Homo erectus 4
Homo sapiens neanderthalenis 3

Homo sapiens sapiens 3
Horgan, John 15
House of Hohenzollern 138
Howe, Gen. William 99

The Iliad 68
Iroquois 116
Islam 193
Isocrates 47, 80, 86

Jaynes, Julian 41
Jefferson, Thomas 80, 81
Jericho 2, 7, 8, 39
Jesus of Nazareth 80
Jung, Carl 62–64

kamikaze 160
Keegan, John 1, 4, 7, 12, 19, 64, 141, 196
Keith, Sir Arthur 17
King George III 81
King Sennacherib 33
Kohlberg, Lawrence 48

Lao Tzu 80
Lenin, Nikolai 166
Lincoln, Abraham 113–136
Lorenz, Konrad 8
Luther, Martin 83–86

Mallory, Stephen 121
Mann, Thomas 20, 55
Mao Tse-Tung 166
Marne River 142
Marx, Karl 164
McCullough, David 93, 97
Mein Kampf 153
Merrimac 120–124
Middle East 194
Mitani, John 13
Mohammed 80
Momaday, N. Scott 29
Monitor 120
Morgan, Daniel 105

Napoleon 70
New Hampshire, "Live Free or Die" 174
New Stone Age 5
Newburgh Conspiracy 108
Newton, Isaac 86
Ngogo chimpanzees 13–16, 58

Oakley, Kenneth 189
O'Connell, Robert 4–13, 19, 35, 47, 48, 57–65, 69, 79, 90, 117, 121, 139, 143, 152–168, 185–187
The Odyssey 68
O'Hara, Gen. Charles 90
Old Stone Age 4
Old Testament 33, 43
Operation Barbarossa 156
Oppenheimer, Robert 162

Pacifism 143–146
Pearl Harbor 88, 156–157
Phipps, Carter 23
Piaget, Jean 23
Plato 47
projection 50–60
Pruyser, Paul 86
Ptolemy 83
Pyrrahus the Epirot 157

Rice, Phillip 23
Robertson, Dr. Robin 63
Romanov Dynasty 138
Rousseau, Jean-Jacques 35, 45
Rubin, Dr. Theodore 54–57

Sabine, George 103
Sagan, Carl 44
Sarissa 72
Schlieffen Plan 139
scientific method 180
Sedgwick, Gen. John 120
sensorimotor 24
Siddhartha Gautama 80
The Social Contract 35
Springfield rifle, 1861 118
Stalin, Joseph 166
Stanton, Edwin 122
Storr, Anthony 49, 50, 52, 67
Stowe, Harriet Beecher 115
Stuart, J.E.B. 48

Tarleton, Col. Banastre 106
Thompson, Clara 53
totemism 30
Trenton, Battle of 100
Trinkaus, Erik 15
Tuchman, Barbara 139
Turney-High, Harry 6

U.S. Constitution, 13th Amendment to 135

warrior kings 67
Washington, George 93–110
Watts, David P. 13
Weimar Republic 104
Welles, Gideon 122
Whitehead, Alfred North 9
Whitney, Eli 118

Wilber, Ken 23–39, 45, 49, 57–68, 85, 103, 114–116, 146, 149, 164–181, 188, 192, 193
Wilson, Edward O. 19
Wilson, Woodrow 149
Winter War 89
Wrangham, Richard 13, 14

Yale 96

www.ingramcontent.com/pod-product-compliance
Lightning Source LLC
Chambersburg PA
CBHW032056300426
44116CB00007B/760